BRITISH GEOLOGICAL SURV
Natural Environment Researc

British Regional Geology

# Central England

THIRD EDITION

B. A. Hains, BSc, PhD, and A. Horton, BSc

*Based on previous editions by*
F. H. Edmunds, MA, and K. P. Oakley, BSc, PhD

LONDON  HER  MAJESTY'S  STATIONERY  OFFICE  1969

**HER MAJESTY'S STATIONERY OFFICE**

HMSO publications are available from:

**HMSO Publications Centre**
(Mail and telephone orders)
PO Box 276, London SW8 5DT
Telephone orders (01) 622 3316
General enquiries (01) 211 5656
*Queueing system in operation for both numbers*

**HMSO Bookshops**
49 High Holborn, London WC1V 6HB
 (01) 211 5656 (Counter service only)
258 Broad Street, Birmingham B1 2HE
 (021) 643 3740
Southey House, 33 Wine Street, Bristol
 BS1 2BQ  (0272) 264306
9 Princess Street, Manchester M60 8AS
 (061) 834 7201
80 Chichester Street, Belfast BT1 4JY
 (0232) 238451
71 Lothian Road, Edinburgh
 EH3 9AZ  (031) 228 4181

**HMSO's Accredited Agents**
(see Yellow Pages)

*And through good booksellers*

**BRITISH GEOLOGICAL SURVEY**
Keyworth, Nottinghamshire NG12 5GG
 Plumtree (060 77) 6111
Murchison House, West Mains Road,
 Edinburgh EH9 3LA  (031) 667 1000

The full range of Survey publications is available through the Sales Desks at Keyworth and Murchison House. Selected items are stocked by the Geological Museum Bookshop, Exhibition Road, London SW7 2DE; all other items may be obtained through the BGS London Information Office in the Geological Museum ((01) 589 4090). All the books are listed in HMSO's Sectional List 45. Maps are listed in the BGS Map Catalogue and Ordnance Survey's Trade Catalogue. They can be bought from Ordnance Survey Agents as well as from BGS.

*The British Geological Survey carries out the geological survey of Great Britain and Northern Ireland (the latter as an agency service for the government of Northern Ireland), and of the surrounding continental shelf, as well as its basic research projects. It also undertakes programmes of British technical aid in geology in developing countries as arranged by the Overseas Development Administration.*

*The British Geological Survey is a component body of the Natural Environment Research Council.*

© *Crown copyright 1969*

*First published 1936*
*Third edition 1969*
*Third impression 1987*

ISBN 0 11 880088 4

*Maps and diagrams in this book use topography based on Ordnance Survey mapping*

# Foreword to Third Edition

The first and second editions of 'The Central England District' were written jointly by the late Mr. F. H. Edmunds and Dr. K. P. Oakley and published in 1936 and 1947 respectively. Since 1947 much new information on the geology of the region has become available and consequently, in the third edition, most of the text has been rewritten and the illustrations replaced or revised. Dr. Hains has revised the chapters dealing with Pre-Cambrian and Palaeozoic rocks and Mr. Horton those dealing with Mesozoic and Quaternary rocks. Many other Geological Survey officers have supplied notes and information, especially Mr. W. B. Evans on the Coal Measures of North Staffordshire (including Fig. 10), Mr. B. J. Taylor on the Triassic of the Cheshire Basin, Dr. A. A. Wilson on the Millstone Grit Series of North Staffordshire (including Fig. 7), Dr. A. W. A. Rushton on the Palaeozoic faunas and Dr. H. C. Ivimey-Cook on the Jurassic faunas. Mr. G. A. Kellaway has contributed an account of the Pleistocene Structures and Quaternary Earth Movements (pp. 114-9) and has acted as Editor.

Institute of Geological Sciences,
Exhibition Road,
South Kensington,                                                       K. C. DUNHAM
London, SW7 2DE                                                *Director*
29th October 1968

*An EXHIBIT illustrating the Geology and Scenery of the region described in this volume is set out on the first gallery of the Museum of Practical Geology, Exhibition Road, South Kensington, London SW7 2DE.*

# Contents

| | | Page |
|---|---|---|
| 1. | **Introduction:** History of Research; Table of Formations | 1 |
| 2. | **Pre-Cambrian:** Uriconian; Charnian | 6 |
| 3. | **Cambrian** | 10 |
| 4. | **Silurian:** Upper Llandovery Series; Wenlock Series; Ludlow Series | 15 |
| 5. | **Old Red Sandstone:** Lower Old Red Sandstone; Upper Old Red Sandstone; The Mountsorrel Granodiorite | 22 |
| 6. | **Carboniferous:** Carboniferous Limestone Series; Millstone Grit Series; Coal Measures | 28 |
| 7. | **Permo Carboniferous** | 57 |
| 8. | **Permo-Triassic:** Permian; Bunter; Keuper; Rhaetic; Mineralization and Intrusive Igneous Rocks | 60 |
| 9. | **Jurassic:** Lias; Inferior Oolite Series; Great Oolite Series and Cornbrash; Kellaways Beds and Oxford Clay | 74 |
| 10. | **Pleistocene and Recent Deposits:** Nomenclature and Classification; Sequence; Periglacial Deposits; Post-Glacial and Recent Deposits; Changes in River Drainage | 89 |
| 11. | **Economic Geology** | 104 |
| 12. | **Structure:** Pre-Old Red Sandstone Movements; Intra-Carboniferous and Permo-Carboniferous Movements; Post-Triassic Movements; Pleistocene Structures; Quaternary Earth Movements | 110 |
| 13. | **Maps and Memoirs of the Geological Survey, and Other References relevant to Central England** | 120 |
| 14. | **Index** | 132 |

# Illustrations

## Figures in Text

| Fig. | | Page |
|---|---|---|
| 1. | Generalized section of the Cambrian rocks of the Nuneaton area | 11 |
| 2. | Cambrian Fossils | 13 |
| 3. | Palaeogeographical sketch-maps illustrating the progressive flooding of the 'Midland Block' by the sea during the Silurian Period | 16 |
| 4. | Silurian Fossils | 19 |
| 5. | Palaeogeographical sketch-map showing the possible area of St. George's Land and the Midland Barrier; the land area (in D-Zone times) is shown stippled | 29 |
| 6. | Limits of the Widmerpool Lower Carboniferous gulf | 33 |
| 7. | The succession and goniatite stages of the Millstone Grit Series of North Staffordshire | 35 |
| 8A. | Block-diagram illustrating the origin of split coal seams | |
| 8B. | Diagrammatic section showing the northward splitting of coal seams in the South Staffordshire Coalfield, and the correlation of seams on the evidence provided by marine bands | 40 |
| 9. | Diagram of a 'wash-out' in the Thick Coal of Warwickshire: Victoria Colliery, Hawkesbury | 42 |
| 10. | Generalized sequence and classification of the Coal Measures of the North Staffordshire (Potteries) Coalfield | 46 |
| 11. | Reconstructed section through the Coalbrookdale Coalfield showing the unconformity termed the 'Symon Fault' between the Upper Coal Measures (Coalport Beds) and the Lower and Middle Coal Measures | 49 |
| 12. | Generalized vertical sections illustrating the lateral and vertical variations of the Permo-Triassic rocks | 61 |
| 13. | Palaeogeography of the Bunter Pebble Beds | 65 |
| 14. | Sketch-map illustrating the existing distribution of the Permo-Triassic rocks and the progressive expansion of the depositional basin | 67 |
| 15. | Generalized section to show changes in the succession from north to south of part of the Middle Jurassic of Central England | 81 |
| 16. | Sketch-map illustrating the location of the Jurassic ironstone fields and limits of workable ironstone | 82 |
| 17. | Structural map of Central England | 112 |

## Tables in Text

| Table | | Page |
|---|---|---|
| 1. | Classification of the Coal Measures | 44 |
| 2. | Regional succession of the Quaternary deposits of Central England and East Anglia | 91 |
| 3. | Generalized Pleistocene sequence and deposits of the Shropshire–Cheshire basin | 93 |

## Plates

| Plate | | | Facing page |
|---|---|---|---|
| I. | | Physiographic Map of Central England | 1 |
| II. | | Sketch Map of the Solid Geology of Central England | 4 |
| III. | | Crags of Beacon Hill Beds, Beacon Hill, Charnwood Forest. Crags of fine-grained, banded, silicified ashes (hornstones) with a strong vertical cleavage. Beyond the crags the ground falls away to the Soar valley, floored with Triassic rocks and alluvium. [SK 509 148]. (Geological Survey Photograph No. A10255) | 18 |
| IV. | | Characteristic Silurian Fossils | 19 |
| V. | A. | Titterstone Clee Hill, Shropshire. A basalt sill caps the summit. Thick screes of basalt debris obscure the highest Ditton Series rocks of the lower slopes. [SO 585 772]. (Geological Survey Photograph No. A9512) | |
| | B. | Downton Castle Sandstone, Onibury, Shropshire. Well developed cross-bedding in yellow, flaggy, micaceous sandstone. [SO 455 793]. (Geological Survey Photograph No. A9528) | 42 |
| VI. | | Characteristic Fossil Plant Remains from the Coal Measures | 43 |
| VII. | | Generalized Sections of the Lower and Middle Coal Measures of the Coalfields of Central England | 52 |
| VIII. | A. | A supposed animal burrow produced by the trace fossil '*Diplocraterion*' in a shell-fragmental limestone of the Great Oolite Limestone. Quarry, south-west of Cosgrove, Northamptonshire. [SP 784 420]. (Geological Survey Photograph No. A10165) | |
| | B. | Keuper Conglomerate on Keuper Passage Beds. The Keuper Conglomerate is the hard overhanging rock. It rests upon the softer thinly bedded Keuper Passage Beds. The outcrop of the Conglomerate forms the summit of a west facing escarpment. Near Muskets Hole [SJ 508 548] on Raw Head, near Bulkeley, Cheshire. (Geological Survey Photograph No. L9) | 68 |

| | | | |
|---|---|---|---|
| IX. | | Characteristic Jurassic Fossils | 84 |
| X. | A. | Upper Lincolnshire Limestone (Weldon Stone). Oolitic freestone quarry, ¼ mile west-south-west of Great Weldon [SP 925 892]. (Geological Survey Photograph No. A8353) | |
| | B. | Glacial Deposits, Shawell Gravel Pit, near Rugby. Upper Chalky Boulder Clay resting on Glacial Sand and Gravel [SP 535 797]. (Geological Survey Photograph No. A10128) | 85 |
| XI. | | The mammoth W1400 dragline excavating deep overburden in a Northampton Sand Ironstone pit. The small excavator is loading the ironstone into the trucks. The team in the centre of the photograph are drilling the ironstone prior to shot-firing. The overburden is worked in two stages. First the Chalky Boulder Clay and Upper Estuarine Series are dug, then the bared surface of the Lincolnshire Limestone is shot-fired before it and the underlying Lower Estuarine Series are removed. (Published by permission of Stewarts and Lloyds Limited) | 106 |

# 1. Introduction

The Central England Region includes the counties of Northamptonshire, Leicestershire, Rutland, Warwickshire and Staffordshire and parts of Derbyshire, Nottinghamshire, Worcestershire, Cheshire, Flintshire and Shropshire. Its area, which is arbitrarily delimited, is shown on the sketch map, Plate I. Most of the region lies below the 400 ft contour, and with but few exceptions the remainder is below 800 ft. The Clent Hills attain a height of 1036 ft (326 m); north of Stoke-on-Trent the ground reaches 1102 ft (336 m) and the Clee Hills rise to 1790 ft (546 m).

The physical features and scenery are intimately related to the character of the underlying rocks. The rocks exposed at the surface are almost wholly of sedimentary character, but igneous rocks of various ages and types occur in some places, notably in Charnwood Forest, at Rowley Regis and near Nuneaton. All geological systems from Pre-Cambrian to Jurassic, with the possible exception of the Ordovician, are represented at the surface (Plate II). The sequence of strata is given in the 'Table of Formations', p. 2–3.

The youngest solid rocks, of Jurassic age, crop out in the eastern and southern parts of the region. The Middle Jurassic sediments give rise to a gently undulating plateau with a deeply indented escarpment at its western extremity. The underlying Lias forms extensive clay vales with minor escarpments and plateaux produced by the outcrop of the Marlstone Rock Bed. Permo-Triassic rocks cover a larger area than any of the other systems. Wide expanses of Keuper Marl generally form uninteresting low-lying country, only broken in places by ridges produced by occasional sandstone beds. The underlying arenaceous strata of Permo-Triassic age produce much more varied scenery with prominent scarplands. Areas formed of Pre-Cambrian and Palaeozoic rocks usually exhibit great variety of surface relief. When these rocks are overlain by or faulted against younger and softer formations the contrast between the landscape formed on the older rocks and that of the Mesozoic terrain is most marked as is well seen in Charnwood Forest (Plate III) and on the north-eastern edge of the Warwickshire Coalfield. The Clee Hills, in the western part of the region, are formed of Old Red Sandstone sediments with a capping of Carboniferous rocks (Plate VA).

In the Cheshire Plain the present day scenery reflects the form and variation of the Drift deposits which cover much of the outcrop of the Permo-Triassic rocks. In the flat region of the western part of the Fens the only features are the sinuous ridges produced by the alluvial sands and silts of the rivers Glen and Welland.

The northern part of the region is drained mainly by the River Trent and its tributaries, and to a lesser extent by the rivers Dee and Weaver; the Severn and the Warwickshire Avon flow through the western and southern parts, while much of the east Midlands is drained by the Welland and the Nene.

## Table of Formations present in Central England

Representatives of the following formations are exposed at the surface:

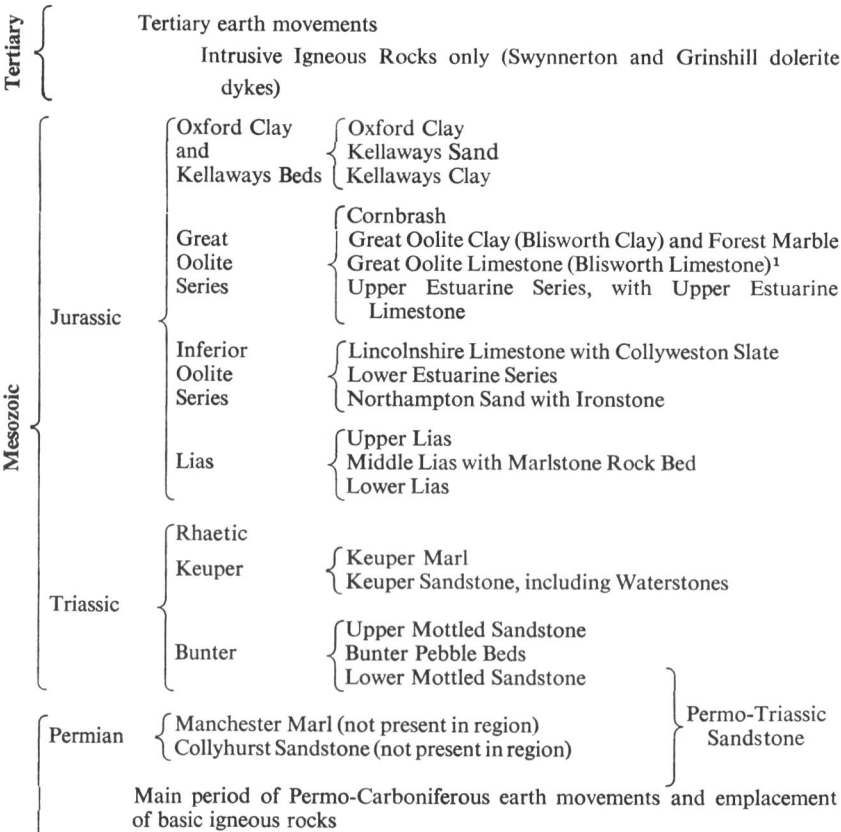

Quaternary
- Recent and Pleistocene: Alluvium; *Chara* Marl, Shell Marl and Tufa; Peat; Fen Deposits; River Gravels; Glacial Sand and Gravel; Glacial Lake Deposits; Boulder Clay; Pre-glacial Sand and Gravel.

Pronounced unconformity following erosion in Tertiary times.

Tertiary
- Tertiary earth movements
- Intrusive Igneous Rocks only (Swynnerton and Grinshill dolerite dykes)

Mesozoic

Jurassic
- Oxford Clay and Kellaways Beds
  - Oxford Clay
  - Kellaways Sand
  - Kellaways Clay
- Great Oolite Series
  - Cornbrash
  - Great Oolite Clay (Blisworth Clay) and Forest Marble
  - Great Oolite Limestone (Blisworth Limestone)[1]
  - Upper Estuarine Series, with Upper Estuarine Limestone
- Inferior Oolite Series
  - Lincolnshire Limestone with Collyweston Slate
  - Lower Estuarine Series
  - Northampton Sand with Ironstone
- Lias
  - Upper Lias
  - Middle Lias with Marlstone Rock Bed
  - Lower Lias

Triassic
- Rhaetic
- Keuper
  - Keuper Marl
  - Keuper Sandstone, including Waterstones
- Bunter
  - Upper Mottled Sandstone
  - Bunter Pebble Beds
  - Lower Mottled Sandstone

Permian
- Manchester Marl (not present in region)
- Collyhurst Sandstone (not present in region)

Permo-Triassic Sandstone

Main period of Permo-Carboniferous earth movements and emplacement of basic igneous rocks

---

[1] In future publications the name Blisworth Limestone will replace Great Oolite Limestone. In north Oxfordshire the latter term includes beds which are equivalent to the Blisworth Limestone ('Great Oolite Limestone') and underlying Upper Estuarine Series of Northamptonshire.

## Introduction

| | | | | |
|---|---|---|---|---|
| **Palaeozoic** | Carboniferous | ? | Enville Beds | |
| | | | Coal Measures | Upper Coal Measures → Keele Group, Newcastle-under-Lyme Group, Etruria Marl Group, Blackband Ironstone Group (and lower beds) |
| | | | | Middle Coal Measures |
| | | | | Lower Coal Measures |
| | | | Millstone Grit Series | |
| | | | Carboniferous Limestone Series | |
| | Old Red Sandstone | | Upper Old Red Sandstone | Farlow Sandstone Series |
| | | | Lower Old Red Sandstone | Clee Group |
| | | | | Ditton Series |
| | | | | Downton Series → Ledbury Group, Temeside Group |

Main period of Caledonian earth movements and probable emplacement of the Mountsorrel intrusion

| | | | | |
|---|---|---|---|---|
| | Silurian | | Ludlow Series | Upper Ludlow Shales, Aymestry (Sedgley) Limestone, Lower Ludlow Shales |
| | | | Wenlock Series | Wenlock (Dudley) Limestone, Wenlock Shales, Woolhope (Barr) Limestone |
| | | | Upper Llandovery Series | Rubery Shale, Rubery Sandstone |

Ordovician(?) Emplacement of post-Tremadoc intrusive igneous rocks

| | | | | |
|---|---|---|---|---|
| | Cambrian | | Stockingford Shales | Merevale Shales (Tremadoc Series), Oldbury Shales, Purley Shales |
| | | | Hartshill Quartzite | Camp Hill Grit, Tuttle Hill Quartzite, Park Hill Quartzite |

Late Pre-Cambrian earth movements and emplacement of the 'syenites' of Charnwood Forest and the 'Blue Hole Intrusive Series' of Nuneaton

| | | |
|---|---|---|
| **Pre-Cambrian** | Charnian of Charnwood Forest | Brand Series, Maplewell Series, Blackbrook Series |
| | Uriconian of West Midlands | Caldecote Volcanic Series (Nuneaton area), Lilleshall and Barnt Green rocks (Coalbrookdale and Lickey Hills) |

The rocks of Central England contain deposits of great economic importance, including coal, fireclay, gypsum, salt and iron ore. As a consequence highly industrialized areas have developed, comprising the Potteries and the Black Country, including Birmingham, Coventry and Stoke-on-Trent. The southern and eastern part of the region is almost wholly of a pastoral character, though the exploitation of important iron-ore deposits in Northamptonshire, Rutland and Leicestershire has been changing the character of some localities.

**History of Research**

Our detailed knowledge of the geology of Central England is of comparatively recent date. Before 1850 little work had been done in the area although several geologists are outstanding for their researches on isolated districts. Robert Plot, in 1686, published a 'Natural History of Staffordshire', in which he described Coal Measures in some detail, and attempted correlation. He described numerous rocks and figured some crystals and fossils. From 1804 John Farey published several geological papers on Derbyshire. In 1808 he drew a section showing the arrangement of strata between Derbyshire and the Lincolnshire coast. William Smith's geological maps were issued in 1815. Strickland covered a wide field in his researches, including Drifts, the Lias and the Triassic, during the twenties of the last century. In 1829 Yates described the Hartshill and Lickey quartzites and the associated intrusive rocks, and concluded that they were of Silurian age.[1]

In 1839 Murchison gave a comprehensive account of the geology of Shropshire, Worcestershire and Staffordshire in his 'Silurian System'. Brodie described fossils from the Lias of the Midlands in several papers and in 1850 discussed the geology of Grantham.

Systematic geological mapping of the region was initiated by the Geological Survey. Central England is a region of much mineral wealth, and as a consequence early Survey memoirs were primarily of an economic nature, dealing in particular with coalfields. Between 1850 and 1880 hand-coloured geological maps covering practically the whole area were published on the scale of one inch to a mile. These 'Old Series' maps have been gradually superseded by 'New Series' maps, on the same scale, but colour-printed. The latter are now being published at the 1:50,000 scale.

A number of Geological Survey memoirs have been produced. Thus in 1859 were published Beete Jukes' account of the South Staffordshire Coalfield, and a similar work by Howell on the Warwickshire Coalfield. In 1860 Hull's 'Geology of the Leicestershire Coalfield' appeared, and during 1886–1891 several volumes by Aveline, with whom Trench and Howell collaborated, were issued; these dealt with the geology of parts of Northamptonshire, Leicestershire, Warwickshire and Nottinghamshire. Since then various economic memoirs have been published dealing with subjects such as Gypsum, Rock Salt, Iron Ore, Coal, Refractory Minerals and Water Supply. In 1900 the first memoir to be associated with a New Series map of part of Central England appeared, viz. 'The Geology of the

---

[1]At that time the term 'Silurian' had a less restricted sense than it has today.

Country between Atherstone and Charnwood Forest', by C. Fox-Strangways and W. W. Watts. Many officers of the Geological Survey have contributed to the various one-inch geological sheets and their explanatory memoirs (Chapter 13). Among the features of Midland geology which are of wider interest and which have been described by past or present members of the Geological Survey are the succession in the Pre-Cambrian rocks of Charnwood Forest and the Pleistocene structures of the Northampton Sand Ironstone Field.

Organized work in Central England by geologists other than those of the Geological Survey staff commenced about 1886. In that year the British Association visited Birmingham, and general interest in the geology of the region was kindled by Charles Lapworth, whose outstanding work forms the foundation of our knowledge of the Pre-Cambrian and Cambrian rocks. Since 1886 much research work has been carried out by the Geological Department of Birmingham University, under the successive direction of Professors Lapworth, W. S. Boulton, L. J. Wills and F. W. Shotton. The numerous papers of Beeby Thompson have added greatly to our knowledge of the Jurassic formations. Other workers who have made valuable contributions to knowledge of the geology of Central England include T. G. Bonney, T. O. Bosworth, E. S. Cobbold, V. C. Illing, W. Wickham King, F. Raw and H. H. Swinnerton.

Ordovician, Cretaceous and Tertiary sedimentary strata are unknown in Central England. Igneous rocks include the Pre-Cambrian 'porphyroids' and 'syenites' of Charnwood Forest, the post-Tremadoc 'diorites' of Nuneaton and the Mountsorrel Granodiorite, intrusive rocks of Carboniferous or Permian age in Leicestershire, South Staffordshire and Shropshire, and extrusive rocks of Carboniferous age in Shropshire. Intrusive rocks of Tertiary age occur in North Staffordshire and Shropshire.

# 2. Pre-Cambrian

Pre-Cambrian rocks occur in a number of small inliers within Central England, and have also been recorded from several boreholes. The rocks have been divided into two groups, the Uriconian and Charnian. The type development of the Uriconian is at The Wrekin in Shropshire, and other rocks also referred to this group form inliers at Nuneaton (Caldecote), Lilleshall and Barnt Green. The exposed Charnian is restricted to the vicinity of Charnwood Forest. Evidence from boreholes at Sproxton (Leicestershire), Great Oxendon and Orton (Northamptonshire), Glinton (near Peterborough) and North Creake (north Norfolk) suggests that a belt of basement rocks, probably of Pre-Cambrian age, extends eastwards from Leicestershire to north Norfolk below a cover of Mesozoic deposits.

The relative ages of the various Pre-Cambrian groups of the Midlands and Welsh Borders are still a matter of controversy. The balance of opinion is that the Longmyndian of Shropshire is younger than the Uriconian and that tuff bands in the Eastern Longmyndian represent the waning stages of the vulcanicity which produced the Uriconian ashes, lavas and intrusions. It is possible that this same volcanic cycle may be represented in the Charnian sequence, with the volcanic rocks of the lower part of the Charnian equivalent to the Uriconian and the higher beds equivalent to the Eastern Longmyndian.

## Uriconian

These rocks are chiefly of volcanic origin, with subsidiary intrusions. The Nuneaton outcrop, the largest area of these rocks in Central England, occupies a strip about 2 miles (3.2 km) long and not more than 300 yards (274 m) wide along the north-eastern side of the Hartshill ridge. Here the Uriconian, overlain unconformably by Cambrian quartzite, has been divided by Allen (1957) into the Caldecote Volcanic Series and the Blue Hole Intrusive Series. The former comprises a group of bedded grey, green and brown coarse crystal-tuffs and fine tuffs with some agglomerates and thin lava-flows, overlain by an unbedded feldspar-quartz-crystal tuff. This unbedded tuff appears to be an 'ignimbrite'—a type of glassy shard tuff deposited by an incandescent ash cloud or *nuée ardente*. The Volcanic Series is intruded by the Intrusive Series, which includes porphyritic basalt and granophyric diorite similar in type to the 'markfieldite' of Charnwood. The beds are also intersected by microdiorite sills, probably of Ordovician age (see p. 14).

At Lilleshall, two miles south of Newport, albite-rhyolites with tuffs and breccias of corresponding composition are seen in a narrow faulted inlier at the northern end of the Coalbrookdale Coalfield. Uriconian rocks have been encountered, below a thin cover of Upper Coal Measures and Triassic, in several boreholes to the north-west of the Coalfield.

## Pre-Cambrian

The distribution of pebbles of Uriconian type in the Enville Beds of Staffordshire and Warwickshire (see p. 59) suggests that Uriconian rocks may underlie a large part of the south-central Midlands. However, the only outcrop of Uriconian in this area is at Barnt Green, at the southern end of the Lickey Hills, where pyroclastic rocks similar to those at Caldecote and Lilleshall form a small inlier. Intrusions of brecciated porphyritic basalt and altered diorite also occur, as at Nuneaton. To the north these rocks are faulted against Cambrian quartzite. None of the Uriconian inliers is large enough to be shown on the sketch-map, Plate II.

Vitric crystal-tuffs, possibly ignimbrites (p. 6), of Uriconian type were encountered at $-1144$ ft ($-349$ m) O.D. at Glinton, near Peterborough. Metamorphosed tuffs and mudstones, similar to the Charnian Swithland Slates, were reached at $-2061$ ft ($-628$ m) O.D. at Sproxton, near Melton Mowbray; welded tuff, also probably Charnian, was proved in boreholes at Great Oxendon and Orton (Market Harborough Memoir (Sheet 170)).

### Charnian

The Pre-Cambrian rocks of Charnwood Forest are seen in a series of isolated outcrops, each surrounded by Triassic beds or Drift, with a total area of some 17 square miles (44 km$^2$). They include both intrusive and extrusive igneous rocks, and a series of sediments largely composed of volcanic ash and other ejected material. Some of the sediments accumulated above sea level, but the majority were formed by the resorting and deposition of volcanic debris in shallow water.

During Permian and Triassic times the Charnian rocks formed hills which were eventually buried by Triassic sediments. This Triassic cover has now been partly removed by erosion, and the summits of the pre-Triassic hills are being revealed. The contrast between the hard Charnian rocks and the softer Triassic sediments gives rise to a diversity of scenery within the Forest.

#### Clastic Rocks

This thick group of rocks, well seen in the eastern, central and southern parts of the Forest, has been subdivided by Watts (1947) as follows:

| | |
|---|---|
| Brand Series | Swithland Slates<br>Trachose Grit and Quartzite<br>Hanging Rocks Conglomerate |
| Maplewell Series | Woodhouse and Bradgate Beds<br>Slate-agglomerate<br>Beacon Hill Beds<br>Felsitic Agglomerate |
| Blackbrook Series | Blackbrook Beds |

The north-western area, from Bardon Hill to Grace Dieu Manor, is more complex, both structurally and stratigraphically. Here, the hornstones (silicified ashes) of the Beacon Hill Beds (Plate III) are replaced by massive agglomerates or 'bomb-rocks', and there are occurrences of intrusive and probably extrusive 'porphyroids' which suggest that this area was a focus of volcanic activity. The sequence outlined above roughly corresponds with the stages of a volcanic cycle. The fine-grained tuffs and hornstones of the Blackbrook Beds indicate a prolonged period of mild volcanic activity, the climax of which is marked by the agglomerates and 'bomb-rocks' of the lower part of the Maplewell Series; thereafter its intensity gradually lessened.

In 1958 two types of fossil, possibly with algal affinities, were described by Ford from the Woodhouse Beds. Both occur as impressions, *Charnia masoni* being a frond-like form, and *Charniodiscus concentricus* a disc-like form.

**Structure of Charnwood Forest**

Towards the close of Pre-Cambrian times the Charnian rocks were folded into a south-eastward plunging anticline now visible in a semi-elliptical series of outcrops, broken by faulting and partly obscured by Triassic and Drift deposits. Numerous fractures occurred which are parallel to the axis of the anticline, the flanks of which were thrust inwards. At a later stage, further fractures developed at right-angles to the axis. The key to the detailed structure of Charnwood Forest is provided by the Felsitic and Slate agglomerates which are hard persistent beds and easily recognizable. Evans (1963) has shown that the cleavage in the Charnian rocks runs obliquely across the main anticline, and suggests that it may have been developed soon after the folding.

**Igneous Rocks**

The oldest igneous rocks of the Charnwood area are the so-called 'porphyroids' which are almost confined to the north-western area; typical examples are seen at Peldar Tor and High Sharpley, near Whitwick, and at Grimley near Thringstone. Several rock-types, including dacites and diorite-porphyries have been included in this group. The term 'porphyroid' is here wrongly applied as the large porphyritic crystals characteristic of most of them are the result of primary crystallization and not of metamorphic recrystallization as the term originally implied. Most of these rocks are now considered to be intrusive; some may be plugs filling old volcanic vents, while others may be lavas. They appear to be approximately contemporaneous with certain of the Charnian agglomerates, since the agglomerates are largely concentrated around the 'porphyroid' masses, and fragments of rock similar to the 'porphyroids' are abundant in the agglomerates. Also, the 'porphyroids' show deformation similar to that of the pyroclastic rocks, and some, as for example at High Sharpley, have been intensely sheared and crushed by late Pre-Cambrian movements.

Intrusive masses of 'syenite' (or 'markfieldite') occupy considerable areas. There are two main groups or types, the southern, comprising the laccolithic intrusions of Markfield, Bradgate and Groby, and a northern and central group mainly with narrow linear outcrops as at Hammercliffe, near Copt Oak, and Bawdon Castle. The southern type is a mottled pink and green holocrystalline rock consisting of red-stained plagioclase-feldspar phenocrysts and green chlorite (replacing hornblende and occasional augite) set in a micrographic intergrowth of quartz and alkali-feldspar. Although syenitic in appearance, the rock is more correctly termed a porphyritic microdiorite. The northern type is predominantly grey and more basic in composition.

The 'syenites' show tectonic structures similar to those of the clastic rocks, so it is probable that they are of Pre-Cambrian age. Supporting evidence is seen at Nuneaton, where boulders of a similar type of 'syenite', derived from an intrusion in the Caldecote Volcanic Series, occur in the basal Cambrian quartzite. Recent age-determinations (Meneisy and Miller 1963) also support a late Pre-Cambrian age for both the 'porphyroids' and 'syenites'.

Isolated masses of igneous rock, surrounded by Keuper, occur to the south of Charnwood Forest, as at Enderby, Croft, Narborough and Sapcote. Apart from the mass at Narborough, which is a dark red porphyritic microdiorite, these intrusions are quartz-diorites or monzonites with some similarity to rocks which occur near the margins of the Mountsorrel Granodiorite (p. 27) at Kinchley and Brazil Wood. The age of these rocks is uncertain.

# 3. Cambrian

In Central England, rocks of Cambrian age are exposed only in small inliers at Nuneaton and Dosthill (south of Tamworth), and presumed Cambrian at Lilleshall and in the Lickey Hills, but evidence from boreholes shows that they have a wide underground distribution. With so few occurrences of Cambrian rocks it is possible to give only a broad picture of the conditions under which they were formed.

This period opened with a marine transgression over a land surface of compacted and folded Pre-Cambrian rocks reduced to low relief by denudation. As the sea moved across the area, the basal conglomerates, quartzites and sandstones of the Lower Cambrian were laid down in the shallow coastal waters. By the beginning of the Middle Cambrian the old land surface was largely submerged, and only fine-grained sediment was brought into most parts of the region. Slow subsidence led to the accumulation of the thick succession of marine shales and mudstones which comprise most of the Middle and Upper Cambrian sequence. At no time, however, does the sea appear to have been deep; breaks in deposition, and the occurrence of sandy, glauconitic and phosphatic layers within the shale sequence, show that the sediments were occasionally affected by wave action or tidal scour, and that subsidence was intermittent. The maximum known thickness of the Cambrian in Central England and the Welsh Borders is about 4000 ft (1220 m).

In contrast with the almost unfossiliferous Pre-Cambrian rocks, the Cambrian sediments, apart from the basal quartzites, contain diverse marine faunas. The system can be divided into four broad units with distinct faunal characteristics: the Lower Cambrian, distinguished by Olenellid trilobites (e.g. *Callavia*); the Middle Cambrian, with the trilobite *Paradoxides;* the Upper Cambrian (excluding the Tremadoc Series) with Olenid trilobites, and the Tremadoc Series with Olenid and other trilobites, some showing Ordovician affinities, as well as dendroid graptolites. A detailed zonal subdivision of the system has been made, based largely on trilobites.

**Nuneaton and Dosthill**

The most important areas of Cambrian rocks in Central England are those adjacent to the Warwickshire Coalfield which may be regarded as an upfaulted block with a general synclinal structure, with the oldest rocks cropping out along the two faulted margins. Inliers of Cambrian rocks occur on the west, at Dosthill, and the east, near Nuneaton.

The Nuneaton inlier includes the largest area of Cambrian rocks in Central England, extending for about nine miles (14 km) from Merevale, near Atherstone, to Bedworth, with a maximum width of about a mile. Unlike the other Cambrian inliers, Lower, Middle and Upper Cambrian rocks are all represented, although the succession (Fig. 1) is attenuated.

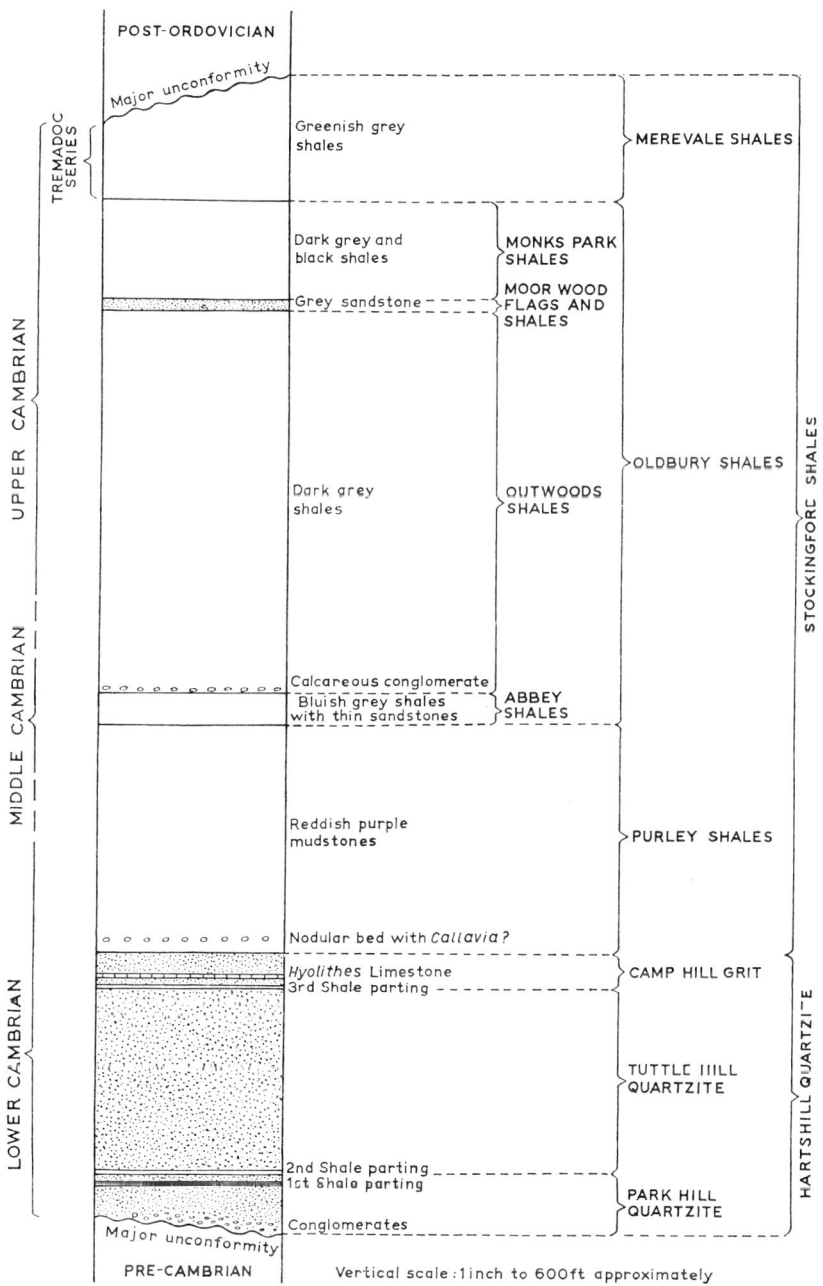

FIG. 1. *Generalized section of the Cambrian rocks of the Nuneaton area*

It was not until Lapworth discovered fossils in the Stockingford Shales in 1882 that the Cambrian age of these beds was established, the unconformity between the Stockingford Shales and the Coal Measures being masked by conformity of strike and similarity in lithology.

The main mass of the Hartshill Quartzite consists of indurated pale pinkish grey sandstones whose siliceous matrix has caused them to assume the characteristics of quartzites. The basal beds are commonly conglomeratic. Beds of purple and grey shale are common, and persistent shale partings have been used to divide the Quartzite into three units (Fig. 1). Near the base of the top division of the Quartzite a thin red sandy limestone occurs. This bed, the *Hyolithes* Limestone, contains pteropod-like molluscs including *Hyolithes* (*Orthotheca*) *degeeri* and horny brachiopods such as *Micromitra* (*Paterina*) cf. *phillipsi*. Fragments of a similar fossiliferous limestone and of quartzite, were found in the Nechells Breccia (p. 57) in a borehole at Nechells, Birmingham.

The Purley Shales are reddish purple mudstones with occasional pale green bands. The occurrence of *Callavia?* (Smith and White 1963, p. 401) in calcareous nodules and shales near the base and of faunas including Eodiscids such as *Serrodiscus* (Fig. 2.1) at higher horizons (Rushton 1966B) indicates a Lower Cambrian age for these beds. Smith and White have obtained an early Middle Cambrian trilobite fauna including *Bailiella emarginata*, *Condylopyge carinata* and *Paradoxides* cf. *sedgwickii* towards the top of the Purley Shales, showing that the bottom of the Middle Cambrian lies within the upper part of the Shales.

The Abbey Shales are bluish grey shales with some thin sandstone bands and contain (Illing 1916) rich Middle Cambrian trilobite faunas which include many species of Agnostids (Fig. 2.2), *Eodiscus punctatus*, *Hartshillia inflata* (Fig. 2.3), *Paradoxides hicksii* (Fig. 2.4), *P. davidis* and *Solenopleura applanata*. A thin conglomerate above this formation marks a non-sequence in the upper part of the Middle Cambrian.

The Outwoods Shales, Moor Wood Flags and Shales and Monks Park Shales have been correlated by Illing (1913) with the Upper Cambrian Maentwrog, Festiniog and Dolgelly groups of North Wales respectively. The Outwoods Shales are mainly dark grey and yield Agnostids such as *Homagnostus obesus*, species of *Olenus* including *O. cataractes* (Fig. 2.6) and, from the upper beds, *Irvingella nuneatonensis* (Fig. 2.5). The overlying Moor Wood Flags and Shales consist mainly of grey fine-grained sandstone, commonly with contorted bedding, and appear to be unfossiliferous. Recent boreholes near Atherstone penetrated the dark grey and black, carbonaceous Monks Park Shales, which contain the brachiopod *Orusia lenticularis* and Olenids including *Parabolina spinulosa* and several species of *Leptoplastus*, *Ctenopyge* and *Sphaerophthalmus* (Rushton 1966A).

The greenish grey Merevale Shales contain the dendroid graptolite *Dictyonema flabelliforme* (Fig. 2.7), typical of the lower part of the Tremadoc Series.

Buff and grey shales crop out in the small inlier at Dosthill. They contain *Hyolithes* and the brachiopod *Acrotreta sagittalis*, and are probably equivalent to part of the Oldbury Shales of Nuneaton.

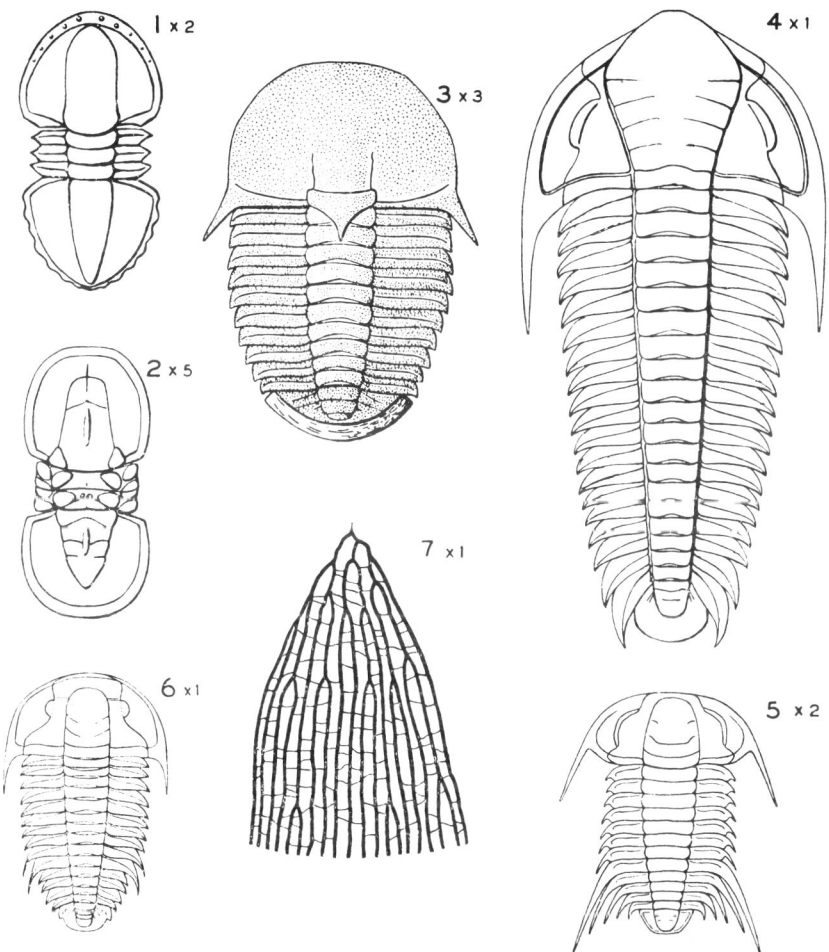

FIG. 2. *Cambrian Fossils*

Purley Shales: 1. *Serrodiscus ctenoa* Rushton; Abbey Shales: 2. *Tomagnostus fissus* (Linnarsson), 3. *Hartshillia inflata* (Hicks), 4. *Paradoxides hicksii* Salter; Outwoods Shales: 5. *Irvingella nuneatonensis* (Sharman), 6. *Olenus cataractes* Salter; Merevale Shales: 7. *Dictyonema flabelliforme* (Eichwald).

## Lickey Hills and Other Areas

A thick mass of greyish white, unfossiliferous quartzite extends for about two miles along the ridge of the Lower Lickey Hills. The quartzite is shattered and faulted, and is folded on north-north-westerly axes. It is thought to be of Lower Cambrian age as it resembles the Hartshill Quartzite of Nuneaton, and is in close association with the Uriconian of Barnt Green (p. 7). White quartzite with green shale partings was proved beneath Upper Llandovery beds at −897 ft (−273 m) O.D. in a boring at Walsall, and quartzite was also reached, below Coal Measures, in a shaft at West Bromwich. A small area of unfossiliferous glauconitic and quartzitic sandstones occurs at Lilleshall. They are probably of Cambrian age as they are lithologically similar to the Comley Sandstone (Lower and Middle Cambrian) of Shropshire, and are adjacent to Uriconian rocks (p. 6).

Cambrian rocks probably form much of the sub-Triassic surface between the Leicestershire and Warwickshire coalfields. Grey shale with *Acrotreta* was recorded below Triassic rocks in a borehole at Shuttington Fields, near Tamworth, and similar, but unfossiliferous, shale in boreholes at Market Bosworth, south-east of Nuneaton and at Sapcote. Black shales of Tremadoc age were proved beneath New Red Sandstone near Leicester, approximately on the line of the Charnwood fold. In the southern part of the Warwickshire Coalfield, Cambrian rocks, mainly shales with occasional sills, have been proved beneath Coal Measures.

**Igneous Rocks in the Cambrian**

In the Nuneaton area and at Dosthill, Cambrian strata have been intruded by numerous sills of dark green hornblendic rocks which range in thickness from a few inches to well over 100 ft (30 m); they have variously been described as diorites, microdiorites and camptonites. Both the quartzites and the shales are intruded and the latter are locally indurated adjacent to the sills. Igneous rocks of similar type invade Cambrian sediments in other parts of the region, and were found in boreholes at Market Bosworth and Leicester, and in several shafts and boreholes in the southern part of the Warwickshire Coalfield. Two drifts at Merry Lees Colliery, on the eastern margin of the Leicestershire Coalfield, passed through Cambrian shales with sills before reaching Coal Measures on the western side of the reversed Thringstone Fault.

These igneous rocks are post-Tremadoc in age and may have been injected during the uplift of the Midland area in Ordovician times—a period of extensive igneous activity in Wales, the Welsh Borders and the Lake District. The intrusions may be associated with the downwarping of the crust to the north and west of Central England which accompanied the development of the Welsh Geosyncline.

# 4. Silurian

No sediments of Ordovician age have been recorded from Central England. The earliest Silurian is also absent, the lowest known beds being of Upper Llandovery Series age. Silurian rocks occur at the surface only in a few small inliers, chiefly within the Coalbrookdale and South Staffordshire coalfields, although they have been proved to underlie part of the West Midlands.

The main lithological subdivisions of these rocks in South Staffordshire and the Lickey Hills are as follows:

|  |  | Approximate thickness ft | (metres) |
|---|---|---|---|
| Ludlow Series | Upper Ludlow Shales | 30 to 50 | (9–15) |
|  | Aymestry (Sedgley) Limestone | 25 | (8) |
|  | Lower Ludlow Shales | 500 | (152) |
| Wenlock Series | Wenlock (Dudley) Limestone | 200 | (61) |
|  | Wenlock Shales | 500 to 700 | (152–213) |
|  | Woolhope (Barr) Limestone | 30 | (9) |
| Upper Llandovery Series | Rubery Shale | 90 | (27) |
|  | Rubery Sandstone | 100 | (30) |

During Ordovician and Silurian times the main tectonic elements which controlled sedimentation in Central England and the adjacent areas were the subsiding Welsh Geosyncline and the relatively stable Midland Block (Wills 1951). The broad boundary zone between these two elements lay in the Welsh Borders. At the close of the Cambrian Period there was widespread uplift over Wales and Central England, with a resulting break in sedimentation over most of the area. In the geosynclinal area the break was short lived, but much of the Midland Block appears to have remained above sea level until Upper Llandovery Series times (Fig. 3). From then onwards there was a contrast in facies between the sediments laid down in the Geosyncline ('basin' facies) and on the Block ('shelf' or 'shelly' facies). In the geosynclinal area subsidence was fairly continuous, and the basin facies comprises shales, greywackes, siltstones and mudstones, with a fauna dominated by graptolites. There was less subsidence on the Midland Block, and the shelf facies consists of a thinner sequence of sandstones, mudstones and limestones with a fauna of brachiopods and other shells. Local intercalations of graptolitic shales occur in rocks of the shelf facies, and aid in the correlation of the two facies.

The basal beds of the Upper Llandovery Series in this region are usually sandstones, laid down as the sea spread over the irregular land surface of the Midland Block. As the land was gradually submerged the sandstones were succeeded by shallow water siltstones and mudstones, and this type of sedimentation continued through most of the rest of the Silurian Period. At times the supply of sediment diminished sufficiently for limestones to be formed, and locally, especially in Wenlock Limestone times, the sea was clear and shallow enough for reef-limestones to develop. An increase in the

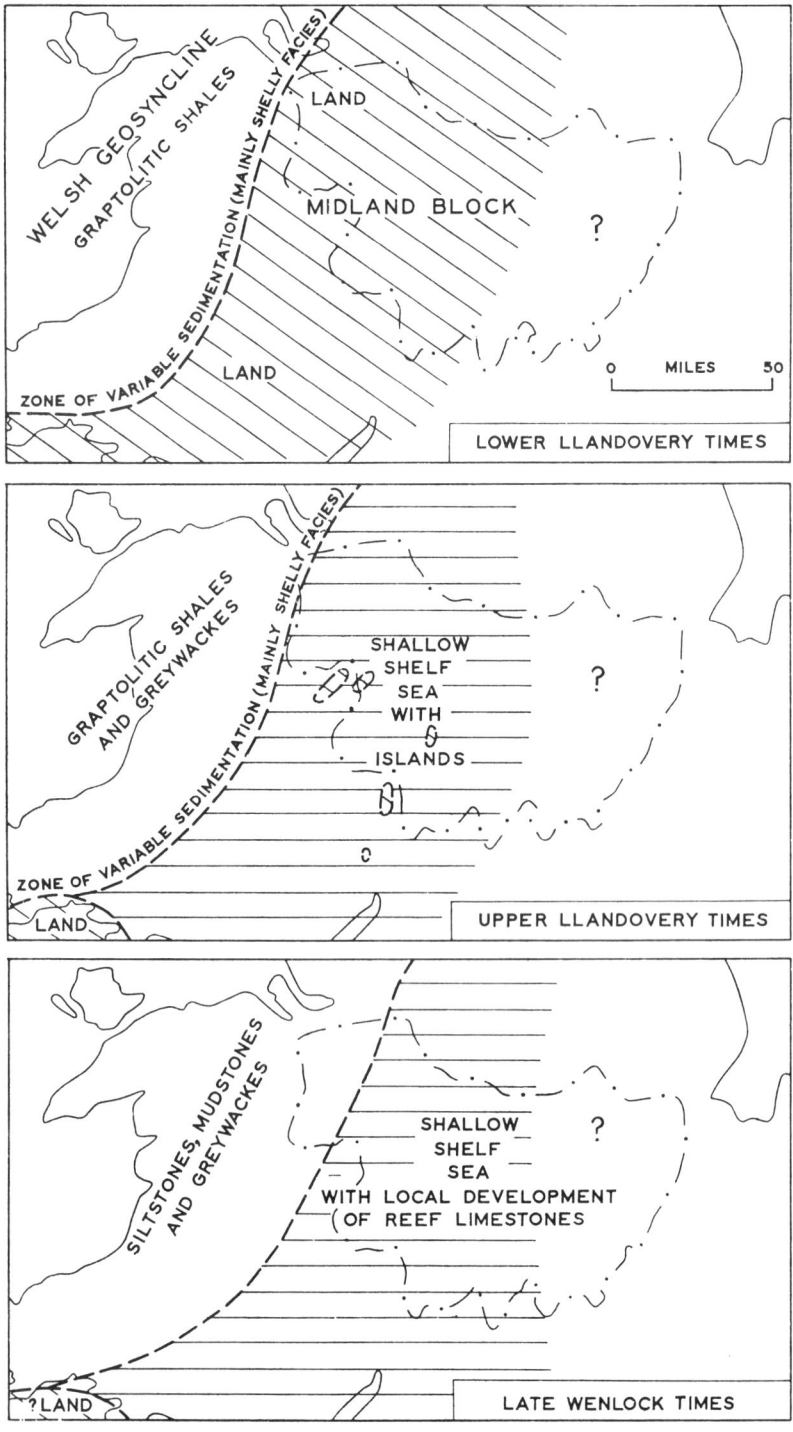

FIG. 3. *Palaeogeographical sketch-maps illustrating the progressive flooding of the 'Midland Block' by the sea during the Silurian Period*

coarseness of the sediment in the higher beds of the Upper Ludlow Shales foreshadowed the change from marine to continental conditions which took place in the Old Red Sandstone Period.

The Silurian rocks which crop out in the classic areas of the Welsh Borders (described in 'The Welsh Borderland', *Brit. Reg. Geol.*) dip south-eastwards under Old Red Sandstone and Coal Measures. They reappear in the Coalbrookdale Coalfield in small inliers, at Barrow, Willey and Linley, on the crests of small anticlines parallel to the strike of the main Shropshire outcrop. They are also exposed in a small faulted anticline, with the same trend, at Neen Sollars, near Cleobury Mortimer. Between the Coalbrookdale and South Staffordshire coalfields Silurian rocks are present beneath Coal Measures and Triassic as proved by a number of boreholes and in South Staffordshire they constitute much of the pre-Carboniferous floor of the Coalfield. These pre-Carboniferous (Silurian and Downton Series) rocks have a gentle regional dip to the north-west. Llandovery rocks crop out on the eastern side of the Coalfield at Great Barr, to the south at Rubery, and appear to be close to the surface near Stourbridge, to the south-west; this distribution suggests that the regional north-westerly dip may represent the eastern limb of a broad syncline. On the eastern side of the Coalfield the Wenlock Series, with a gentle north-westerly dip, is exposed in an inlier east of Walsall, and has also been proved in boreholes and shafts between Walsall and Rowley Regis along the same line of strike. In the centre of the Coalfield the Coal Measures rest on the Ludlow Series, and to the west on the Downton Series. This essentially simple picture is complicated by a series of folds with a north-westerly trend. Two north-westerly plunging synclines are separated by the complex Sedgley and Dudley Anticline. Along this structure three faulted whale-back folds with a north-north-westerly trend, arranged *en échelon*, give rise to inliers of Wenlock and Ludlow beds which form the steep-sided Dudley Castle Hill, Wren's Nest Hill and Hurst Hill. At Sedgley, Ludlow and Downton rocks are exposed in a plunging syncline complementary to the Hurst Hill fold. South-east of Dudley the Sedgley and Dudley Anticline is replaced by a fault which extends southwards to the Lickey Hills, where Llandovery and Wenlock rocks crop out. Ludlow and Downton beds are exposed at Lye and Netherton in small faulted anticlines trending north-north-eastwards. To the east of the Coalfield a borehole two miles east of Great Barr has proved Silurian rocks at −2508 ft (−764 m) O.D.

## Upper Llandovery Series

Beds of this age form small faulted inliers at Rubery (in the northern part of the Lickey Hills), and on the line of the eastern boundary fault of the South Staffordshire Coalfield at Great Barr. They have also been recorded below Wenlock beds in a borehole at Walsall.

Wills (1925) divided the Upper Llandovery Series at Rubery into the Rubery Sandstone (about 100 ft (30 m) thick) below and the Rubery Shale (about 90 ft (27 m) thick) above. The Sandstone, resting unconformably on ?Cambrian quartzite, comprises a basal conglomerate overlain by coarse sandstones with shale beds in the upper part. The Rubery Shale contains thin limestones and sandstones. The fauna consists of numerous brachiopods, including *Atrypa reticularis*, '*Camarotoechia*' *sp.*, *Coelospira hemisphaerica*, *Leptostrophia compressa* and *Stricklandia lirata;* trilobites, *Encrinurus punctatus* and *Phacops elliptifrons;* corals of Streptelasmid type, and the graptolites *Monograptus marri* and *M. nudus*.

The Upper Llandovery Series at Great Barr is represented by fine-grained yellow sandstones with shaly partings and occasional fossiliferous calcareous beds. In the boring at Walsall it consists almost entirely of shales, again resting unconformably on quartzite. An abundance of pebbles of Llandovery sandstone in Midland breccias and conglomerates of Upper Coal Measures and Triassic age indicates that it cropped out over a wide area during these periods.

## Wenlock Series

*Barr Limestone*. The basal member of the Wenlock Series in South Staffordshire is the Barr Limestone, equivalent to the Woolhope Limestone of the Welsh Borders. It forms a narrow outcrop about a mile and a half long, north of Great Barr. The Limestone comprises a group of alternating limestones and shales about 30 ft (9 m) thick with a fauna including the trilobite *Bumastus barriensis* (Pl. IV, fig. 6); the coral *Heliolites interstinctus;* numerous brachiopods, such as *Atrypa reticularis* and *Plectodonta;* sporadic gastropods and the cephalopod *Dawsonoceras annulatum* (Pl. IV, fig. 7).

A small faulted area of limestone, probably the Barr Limestone, occurs at Kendal End, near Barnt Green, surrounded by Pre-Cambrian volcanic rocks. It has also been proved in a borehole at Walsall and a shaft at West Bromwich.

*Wenlock Shales*. This formation comprises greyish green shales with thin limestone beds, from 500 to 700 ft (152–213 m) thick. They are seen only in the South Staffordshire Coalfield, the largest area of outcrop being east of Walsall. They have been proved in several shafts and borings on the south-eastern side of the Coalfield between Halesowen and Walsall.

Fossils are usually abundant and include trilobites, *Dalmanites caudatus* and *Encrinurus punctatus;* corals, *Favosites gothlandicus* and *Heliolites interstinctus* (Pl. IV, fig. 4); brachiopods, *Atrypa reticularis* (Fig. 4.4), *Leptaena rhomboidalis* (Fig. 4.1), *Meristina obtusa*, *Sphaerirhynchia* [*Wilsonia*] *wilsoni* (Fig. 4.2) and *Strophonella euglypha;* lamellibranchs; and gastropods. Graptolites (e.g. *Monograptus priodon*) have been recorded from boreholes at Mucklow Hill, near Halesowen, and at Walsall.

*Wenlock or Dudley Limestone*. The Wenlock Limestone of Staffordshire, known locally as the Dudley Limestone, crops out along the north-western border of the Walsall inlier. It is also seen in the three denuded and faulted folds of Wren's Nest Hill, Dudley Castle and Hurst Hill and has been proved below Coal Measures in collieries between Walsall and Rowley Regis.

In Staffordshire the Wenlock Limestone comprises two limestones separated by the Nodular Beds which consist of shales with limestone nodules and thin limestone bands. The approximate thicknesses of the subdivisions at Dudley are (Butler 1939):

Plate III

Crags of
Beacon Hill Beds,
Beacon Hill,
Charnwood Forest

*For full explanation
see p. viii)*

Plate IV    Characteristic Silurian Fossils

1. *Calymene blumenbachii* Auctt.—Wenlock Limestone. 2. *Gissocrinus luculentus* Ramsbottom—Wenlock Limestone. 3. *Poleumita discors* (J. Sowerby)—Wenlock Limestone. 4. *Heliolites interstinctus* (Linné), var. *decipiens* (McCoy)—Wenlock Limestone. 5. *Acervularia ananas* (Linné), var. *singularis* Lang and S. Smith—Wenlock Limestone. 6. *Bumastus barriensis* Murchison—Barr Limestone. 7. *Dawsonoceras annulatum* (J. Sowerby)—Barr Limestone. 8, 9. *Dayia navicula* (J. de C. Sowerby)—Aymestry Limestone. 10. *Protochonetes ludloviensis* Muir-Wood—Upper Ludlow Shales.

|  |  | ft | (metres) |
|---|---|---|---|
| Upper Wenlock Limestone | { Passage Beds | 9 | (2·7) |
|  | { Upper Quarried Limestone | 25 | (7·6) |
| Nodular Beds |  | 123 | (37) |
|  | { Lower Quarried Limestone | 32 | (10) |
| Lower Wenlock Limestone | { Basement Beds | 10 | (3) |

In the Walsall inlier the Upper Wenlock Limestone is about 12 ft (3·7 m) thick.

A remarkable feature of the Wenlock Limestone is the occurrence of lenses or hemispherical masses of unstratified limestone in the midst of the bedded limestone and shales. The masses have various local names, being known in Staffordshire as 'crog-balls' or 'self-lumps'. They are reef-limestones consisting largely of the skeletons of corals and stromatoporoids, commonly in the position of growth, set in a compact matrix of calcareous mud. Crog-balls vary in size from less than 3 ft (1 m) in diameter to masses 130 ft (40 m) across and upwards of 20 ft (6 m) thick. Surrounding strata arch over, and sometimes sag beneath the crog-balls, probably as a result of differential compaction of the crog-ball and adjacent beds. Crog-balls are thought to have existed as mound-like structures on the sea-floor formed by the prolific growth, at certain centres, of organisms with calcareous skeletons. They were, in fact, diminutive reefs, formed in warm shallow water. The reef-masses are commonest in the Lower Wenlock Limestone and Nodular Beds.

The Wenlock Limestone of Staffordshire contains a rich and varied suite of fossils which differs from that of the Welsh Borders in having a greater abundance of trilobites, polyzoa and well-preserved crinoids. Dudley has long been famous for the beautifully preserved fossils which were obtained when the limestone was quarried, and the commonest Dudley trilobite, *Calymene blumenbachii* (Pl. IV, fig. 1), popularly known as the 'Dudley Locust', attracted sufficient attention in the past to become a local emblem. Other fossils include trilobites such as *Dalmanites;* a rich brachiopod fauna including Orthids (Fig. 4.3), Rhynchonellids, Spiriferids and Strophomenids; corals, including *Acervularia* (Pl. IV, fig. 5), *Coenites*, *Favosites*, *Halysites* ('chain coral') and *Heliolites;* stromatoporoids, *Labechia* and *Stromatopora;* gastropods, including *Poleumita discors* (Pl. IV, fig. 3);

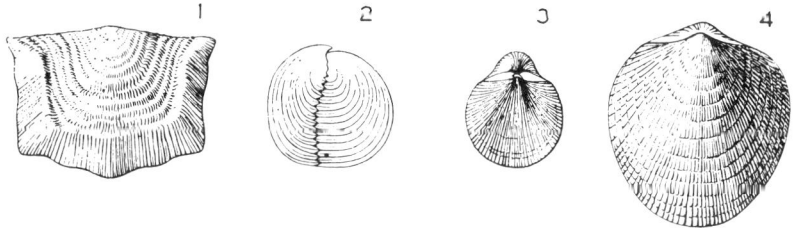

FIG. 4. *Silurian Fossils*

1. *Leptaena rhomboidalis* (Wilckens), 2. *Sphaerirhynchia* [*Wilsonia*] *wilsoni* (J. Sowerby) (side view), 3. *Resserella* [*Parmorthis*] *elegantula* (Dalman), 4. *Atrypa reticularis* (Linné). (All natural size)

lamellibranchs; polyzoa, including species of *Fenestella* and curious pearl-bearing species of *Favositella*, and crinoids (or 'stone lilies') (Pl. IV, fig. 2; Front Cover). Extensive collections of Wenlock Limestone crinoids are exhibited in the Dudley Museum and in the Museum of the Geological Department of Birmingham University.

## Ludlow Series

In the South Staffordshire Coalfield, rocks of the Ludlow Series appear in the Silurian inliers of Dudley, Sedgley, Saltwells, Turner's Hill and Walsall; in the Coalbrookdale Coalfield at Linley, Willey and Barrow, and to the west of the Forest of Wyre at Neen Sollars. They have been proved to underlie the Coal Measures in the Tipton–Oldbury area, and also near Bridgnorth.

*Lower Ludlow Shales.* These beds crop out around Dudley Castle Hill, Wren's Nest Hill, to the north and east of Sedgley, and farther south at Lye and Wollescote and to the west of the Wyre Forest at Neen Sollars. They comprise grey, buff-weathering shales, and sandy mudstones, with thin impersistent limestones and calcareous nodules. They resemble the Wenlock Shales, and both are referred to locally as 'bavin measures', bavin being a local term for the thin limestones and calcareous nodules. The Shales are richly fossiliferous, and characteristic fossils include trilobites, e.g. *Dalmanites nexilis;* cephalopods, e.g. *Gomphoceras ellipticum;* lamellibranchs, and the leaf-like polyzoan *Ptilodictya lanceolata*. Graptolites are rare.

*Aymestry or Sedgley Limestone.* The Aymestry Limestone occurs in several inliers in the Coalbrookdale Coalfield and also in the inlier near Neen Sollars. In the Coalbrookdale Coalfield the Limestone (or the Aymestry Group) is 80 to 100 ft (24–30 m) thick, and consists mainly of calcareous mudstone rather than limestone. The fauna includes the brachiopods *Atrypa reticularis, Leptostrophia filosa, Sphaerirhynchia* [*Wilsonia*] *wilsoni* and *Strophonella euglypha*, and the corals *Favosites gothlandicus forbesi* and *Phaulactis* [*Cyathophyllum*] cf. *angusta. Dayia navicula* (Pl. IV, figs. 8, 9) occurs in the upper part of these beds.

The Sedgley Limestone of South Staffordshire is approximately equivalent to the Aymestry Limestone of the Welsh Borders. North of Sedgley the Silurian rocks are folded into a shallow southerly plunging syncline, and the Sedgley Limestone forms a strong curved escarpment with its highest point at Sedgley Beacon. Farther north, at Park Hill, it forms part of a small faulted dome. South of Sedgley the Limestone has a broad outcrop in the core of a southerly trending anticline. There are small outcrops at Turner's Hill, and in the Netherton Anticline. It consists of about 25 ft (7·6 m) of flaggy and nodular argillaceous limestone. A thin conglomeratic bed at the junction of the Sedgley Limestone and Upper Ludlow Shales at Lye (Squirrell 1958) has yielded thelodont denticles, and a thin impersistent bone bed, also with thelodont denticles, occurs at a comparable horizon at Turner's Hill (Ball 1951).

*Upper Ludlow Shales.* These beds comprise sandy calcareous flags with limestone nodules, silty shale and sandstone. In South Staffordshire, where they crop out at Sedgley, Turner's Hill, Lye and Saltwells, they vary from 30 to 50 ft (9–15 m) thick. The shales are about 100 ft (30 m) thick in the Coalbrookdale inliers and at Neen Sollars.

The fauna of the Upper Ludlow Shales includes the brachiopods '*Camarotoechia*' *nucula*, *Orbiculoidea rugata* and *Protochonetes ludloviensis* (Pl. IV, fig. 10); the lamellibranchs *Fuchsella* [*Orthonota*] *amygdalina* and *Pteritonella* [*Pterinea*] cf. *retroflexa* and the cephalopod *Michelinoceras bullatum*.

## 5. Old Red Sandstone

The main outcrop of the Old Red Sandstone in Central England is located around the Clee Hills in Shropshire. Rocks of this age underlie the Wyre Forest Coalfield to the east, and form the Trimpley and Heightington inliers on the eastern side of the Coalfield. They are also exposed in small inliers in the southern part of the South Staffordshire Coalfield and in the Warwickshire Coalfield near Atherstone and have been recorded from boreholes at Whittington Heath, near Lichfield, and Gayton in Northamptonshire.

The main lithological divisions of the Old Red Sandstone, based on recent work (Ball, Dineley and White 1961; Church Stretton Memoir (Sheet 166)) in the Clee Hills area, are as follows:

|  | Approximate thickness ft | (metres) |
|---|---|---|
| **UPPER OLD RED SANDSTONE** | | |
| Farlow Sandstone Series | up to 500 | (152) |
| Unconformity | | |
| **LOWER OLD RED SANDSTONE** | | |
| Clee Group | up to 900 | (274) |
| Ditton Series (with the Abdon Limestones at the top and the '*Psammosteus*' Limestones Group at the base) | 2000 | (610) |
| Downton Series — Red Downton or Ledbury Group | 1500 | (457) |
| Grey Downton or Temeside Group — Temeside Shales | 40–150 | (12–46) |
| Downton Castle Sandstone | 20–50 | (6–15) |
| Ludlow Bone Bed | up to 1¼ | (0·38) |

The end of the Silurian Period saw the commencement of the main phase of the Caledonian mountain-building movements which raised the sediments of the Welsh Geosyncline into a folded mountain chain. Although the folding did not greatly affect Central England until deposition of the Old Red Sandstone was well advanced it had a profound effect on sedimentation on the Midland Block. The shallow shelf sea of Upper Ludlow times was replaced by a vast slowly subsiding delta-plain fed with sediment from the rising Caledonian mountains, and there was a gradual transition from a marine to a continental environment.

In Central England the earliest indication of the oncoming earth-movements is the increasing coarseness of sediment and the associated impoverishment of the marine fauna towards the top of the Upper Ludlow Shales. The Ludlow Bone Bed marks a pause in sedimentation, when fish remains and other debris were concentrated by winnowing and current action on the floor of a shallow sea. The first signs of the delta-plain which dominated the palaeogeography of Central England and the Welsh Borders during Lower Old Red Sandstone times are seen in the Downton Castle Sandstone. The lower part of the Sandstone resembles the Upper Ludlow Shales in lithology, but the marine shelly fauna of the Shales is replaced by one which is dominated by horny brachiopods and fish. These sediments, probably laid down offshore under brackish-water conditions, are overlain by the deltaic sandstones of the higher part of the Sandstone. This sandy facies is largely confined to the north-western part of the Midland Block. The Temeside Shales show a temporary return to the depositional conditions of the lower part of the Downton Castle Sandstone. The conditions of deposition of the Red Downton Group are still under debate (Allen and Tarlo 1963; Ball, Dineley and White 1961). It is clear, however, that the general transition from a marine to a continental environment was continuing and that the sediments composing this formation were laid down in a deltaic environment at or near sea level. Local marine incursions are marked by the occurrence of bands of molluscs, but it is uncertain whether the main mass of the sediments was laid down in the tidal zone, or on the sub-aerial part of the delta. The sediments are largely siltstones ('marls') with lenticular sandstone beds, and rhythmic bedding is developed, though to a lesser extent than in the Ditton Series (see below). The red colour of the siltstones may be due to derivation from an area of tropical lateritic weathering on the rising mountain chain to the west and north-west.

Apart from minor marine incursions in the lower part, the sediments of the Ditton Series are entirely of continental origin. They comprise a rhythmic repetition of a sedimentary succession consisting ideally of conglomerate, overlain by sandstone, siltstone and occasionally, limestone, resulting from the periodic flooding of an extensive delta. Each flood caused the breaking up of some of the deposits previously laid down on the delta top, thus forming an intraformational conglomerate (conglomeratic cornstone). These conglomerates commonly contain a high proportion of fragments of concretionary limestone, formed during the final stage of the preceding rhythm. As the force of the flood declined, the coarse detritus now forming the conglomerates was buried by sand and then by silt. In many cases there was sufficient desiccation to cause deposition of chemically precipitated concretionary limestones (cornstones) from the sheets of water left by the floods. Most of these limestones were broken up by the next flood, but in the '*Psammosteus*' Limestones Group and the Abdon Limestones they are commonly preserved intact. Rhythmic sedimentation continued in Clee Group times, but there was an increasing proportion of coarse-grained sandstones laid down as the mountain-building movements progressively affected areas nearer to Shropshire. Ultimately the movements led to the uplift and erosion of the whole Clee Hills area prior to the deposition of the Farlow Sandstone Series (Upper Old Red Sandstone) which shows a return to sedimentation on a sub-aerial delta; the closing stages were marked by the gradual transgression of the Lower Carboniferous sea.

As the Downton Series is in many respects transitional between the marine Silurian and the continental Ditton Series, some authorities have included it, or some portion of it, in the Silurian System while others have regarded it as part of the Old Red Sandstone. In recent years the boundary has usually been taken at the base of the Ludlow Bone Bed (White 1950), but Tarlo (1964), in common with many European workers, prefers to include the Downton Series within the Silurian. In the Anglo-Welsh area there is a marked change of fauna at the Ludlow Bone Bed, with the virtual elimination of the Silurian marine fossils and the incoming of the first extensive fish fauna. Fish are the most significant fossils in the Old Red Sandstone, occurring mainly as fragments in the conglomeratic cornstones. Much of our knowledge of the stratigraphical palaeontology of the Old Red Sandstone of the Anglo-Welsh area is due to the work of King (1924, 1925, 1934), who proposed a series of subdivisions based partly on the fish faunas and partly on lithology. Later palaeontological work, chiefly by White (1946, 1950), has led to the formulation of a zonal sequence which has enabled a partial correlation with the Old Red Sandstone of the Continent.

## Lower Old Red Sandstone
### Clee Hills District

Rocks of this age crop out over a wide area around the Clee Hills where there is a complete sequence from the Ludlow Bone Bed up to the Clee Group.

The steep upper slopes of Brown Clee Hill are formed by beds of the Clee Group capped by small unconformable outliers of Coal Measures. Below these slopes, a relatively level platform is formed by the Ditton Series. The sandstones immediately above the '*Psammosteus*' Limestones give rise to the prominent scarp at the margin of this platform, while the relatively soft beds of the Downton Series form the lower slopes of this scarp and occupy the low ground around Acton Round, along Corve Dale and to the south of Titterstone Clee Hill.

*Downton Series.* The basal member of the Downton Series, the Ludlow Bone Bed, may occur as a single thin bed or as a series of closely spaced lenticular beds. It is a brown sandy mudstone or locally calcareous sandstone with abundant fish debris and phosphatic concretions. The commonest fish remains are thelodont and acanthodian denticles and fin spines of *Onchus*. Other fossils include ostracods, fragments of Eurypterids and occasional brachiopods such as *Lingula* and *Protochonetes*.

The basal few feet of the overlying Downton Castle Sandstone lithologically resemble the Upper Ludlow Shales, but the main part is a fine-grained, commonly false-bedded, micaceous yellow sandstone (Plate VB). The succeeding Temeside Shales are in some respects transitional between the Sandstone and the Red Downton Group, and comprise green and purple mudstones and siltstones, with beds of yellowish green, micaceous, flaggy sandstone. The faunas of the Downton Castle Sandstone and Temeside Shales are similar in most respects, and include *Lingula minima* (characteristic of the Downton Castle Sandstone) and *L. cornea* (characteristic of the Temeside Shales and higher Downton beds), lamellibranchs such as *Modiolopsis complanata*, gastropods, ostracods, eurypterid fragments and fish remains. The problematical spherical body *Pachytheca* and fragments of the early land plants *Cooksonia* and *Prototaxites* are common.

The Red Downton Group is composed mainly of red and purple siltstones ('marls') with thin lenses of purple micaceous sandstone. Some cornstone nodules and conglomeratic cornstone beds occur near the top of the Group. The fish fauna includes *Ischnacanthus wickhami* and *Traquairaspis symondsi*, but fossils are uncommon, especially in the lower part of the Group.

*Ditton Series*. The lower boundary of the succeeding Ditton Series is usually taken at the base of the '*Psammosteus*' Limestones Group. The Series is distinguished lithologically from the Downton Series by a higher proportion of sandstones and conglomeratic cornstones. There is also a faunal change at about this horizon, the *Traquairaspis* fauna being replaced by one characterized by *Pteraspis*.

The '*Psammosteus*' Limestones Group comprises several repetitions of a sedimentary rhythm (p. 23) which starts with a conglomeratic cornstone, succeeded by fine-grained brown sandstones with the main part of the sequence composed of red or brown siltstones. These may contain calcareous concretions near the top which grade into lenticular concretionary limestones (cornstones) up to 15 ft (4·5 m) thick. The conglomeratic cornstones contain a fish fauna including *Pteraspis* (*Simopteraspis*) *leathensis*, *P. rostrata* and *Anglaspis*. The fish *Traquairaspis symondsi* (formerly identified as *Psammosteus anglicus*) occurs most commonly in the upper part of the Downton Series.

Similar rhythmic sedimentation persists through the remainder of the Ditton Series, but the cornstone phase is rarely developed except for the Abdon Limestones at the top of the sequence. Massive sandstones are developed above the '*Psammosteus*' Limestones Group, and these give rise to the marked scarp above Corve Dale. The Abdon Limestones are two cornstone horizons well developed above Abdon, on the north-west slopes of Brown Clee Hill. Although fossils are sporadic above the '*Psammosteus*' Limestones Group, the fish faunas include *Pteraspis* (*Belgicaspis*) *crouchi* and *P.* (*Cymripteraspis*) *leachi*.

*Clee Group*. This Group consists of predominantly coarse-grained, often pebbly, green and brown sandstones with subordinate red siltstones in the lower part. These beds are unfossiliferous, but may be approximately equivalent to the Senni Beds and part of the Brownstones of South Wales.

**Trimpley and Heightington**

The Trimpley and Heightington inliers are on the northward extension of the Malvern–Abberley Axis, on the eastern side of the Wyre Forest–Mamble Coalfield.

The Trimpley inlier is structurally complex, but consists essentially of a northern synclinal area and a southern anticlinal area separated by an east west fault. The sequence includes approximately the top 450 ft (137 m) of the Downton Series and about 1200 ft (366 m) of the overlying Ditton Series; the lowest beds of the former are poorly exposed red siltstones. Higher in the sequence, in Man Brook, near Trimpley, green siltstones yield the lamellibranchs *Grammysia? anceps* and *Modiolopsis complanata trimpleyensis*, Eurypterids and fish remains. The succeeding strata are mainly red siltstones with thin sandstone and calcareous horizons. King (1924) called the lower part of these beds the Trimpley Fish Zone (*Ischnacanthus* Zone), about 1400 ft (427 m) above the Ludlow Bone Bed, with fish including *Didymaspis grindrodi* and *Ischnacanthus wickhami*. The higher beds contain the ostracod *Leperditia*. The '*Psammosteus*' (or Birch Hill) Limestones Group is about 100 ft thick, and has the same lithological characteristics as in the Clee Hills area. The remainder of the Ditton Series consists predominantly of siltstones, although sandstones and conglomeratic cornstones are more common than in the Downton Series. The Eurypterid Sandstones of King (1934), immediately above the '*Psammosteus*' Limestones Group, contain the fish *Macropetalichthys* and *Onchus*, and numerous Eurypterids such as *Pterygotus* and *Stylonurus symondsi*.

The Heightington inlier lies to the west of Stourport and to the north of the Abberley Hills. There is probably a full sequence from the Ludlow Bone Bed to high in the Ditton Series, but this is not certain owing to the structural complexity of the area which is affected by a number of powerful northerly trending faults and associated folds. Greenish yellow sandstones, probably equivalent to the Downton Castle Sandstone, crop out north of Abberley Hill, and the upper part of the Downton Series, composed of red siltstones with some sandstone beds, is well exposed in Dick Brook, south of Heightington. The '*Psammosteus*' Limestones Group occurs along the southern margin of the high ground around Heightington and Rock. Near Rock these beds yield a fauna including *Anglaspis macculloughi, Pteraspis (Simopteraspis) leathensis* and *Thelodus*. Higher beds contain the lamellibranchs *Modiolopsis* and *Vigorniella regis*. The upper part of the Ditton Series is poorly exposed.

Beds of Lower Old Red Sandstone age underlie the Wyre Forest–Mamble Coalfield and occur locally below Coal Measures in the southern part of the Coalbrookdale Coalfield.

**South Staffordshire and Warwickshire**

Downton Series beds form the sub-Coal Measures floor in the north-western and south-western parts of the South Staffordshire Coalfield (p. 48), and are exposed in a number of small inliers in the south-western part. The largest inlier is at Turner's Hill and Gornal, south of Sedgley, where Silurian and Lower Old Red Sandstone strata occur in three small anticlinal areas, separated by faults. The Ludlow Bone Bed, up to 15 inches (38 cm) thick, is succeeded by about 30 ft (9 m) of buff silty sandstones, shales and mudstones, known as the Turner's Hill Beds. Although similar in lithology to the Upper Ludlow Shales, the Ludlow fauna is absent and they contain *Lingula cornea* and Eurypterids. The beds are overlain by the Gornal Sandstone, 35 ft (11 m) of yellow-buff massive sandstone which was originally thought to be within the basal part of the Coal Measures. The discovery of the fish *Hemicyclaspis murchisoni* and other fossils proves it to belong to the Downton Series and together with the Turner's Hill Beds it is now considered to be equivalent to the Downton Castle Sandstone of the Welsh Borders (Ball 1951). Above the Gornal Sandstone are about 25 ft (7·6 m) of Temeside Shales, overlain by the highest beds in the inlier, the lower part of the Red Downton Group, which are purple siltstones with purple and green micaceous sandstones containing the fish *Didymaspis, Hemicyclaspis* and *Onchus*. South-west of Netherton, a small inlier in the core of the Netherton Anticline shows a sequence from Upper Ludlow Shales to the Red Downton Group. Here, the Temeside Shales, purple and olive-green siltstones and mudstones with some sandstones and thin bone beds, are well exposed at Brewin's Bridge. The Netherton axis also brings up Downton Series strata at Lye and Wollescote, and there is a small area of the Red Downton Group at Coalbournbrook, on the western edge of the Coalfield.

## Upper Old Red Sandstone

*Farlow Sandstone Series.* This Series crops out in a limited area along the north-eastern flank of Titterstone Clee Hill. It rests unconformably on gently folded Lower Old Red Sandstone strata, and is overstepped by the basal Lower Carboniferous. The lower part of the Series consists chiefly of fine-grained yellowish false-bedded sandstones with many pebbly beds, and a basal conglomerate up to 10 ft (3 m) thick. These beds are overlain by a succession of grey, green and purple calcareous sandstones, locally pebbly, with occasional siltstones and thin rubbly limestones. The lower beds contain a fish-fauna, distinct from that of the Lower Old Red Sandstone, including species of *Bothriolepis, Pseudosauripterus* and *Eusthenopteron.*

A boring at Whittington Heath (Fig. 5, No. 6,) between Lichfield and Tamworth, entered beds of Upper Old Red Sandstone age at 3670 ft ( 1119 m) O.D., beneath thin Carboniferous Limestone. Unfossiliferous coarse red sandstones with sporadic pebbles, similar to some beds in the Farlow Sandstone Series of the Clee Hills, were proved below Carboniferous Limestone in a borehole at Gayton, Northamptonshire (Fig. 5, No. 9).

Taylor (1966) has shown that Old Red Sandstone is present on the eastern flank of the Warwickshire Coalfield. Here, greenish grey sandstones, formerly thought to be basal Coal Measures, contain a fauna indicative of the Upper Old Red Sandstone including *Bothriolepis* and *Holoptychius, Lingula cornea* and other brachiopods, lamellibranchs and polyzoa.

## The Mountsorrel Granodiorite

An intrusive boss of granodiorite and associated rocks, more than a mile in diameter, occurs at Mountsorrel, east of Charnwood Forest. The granodiorite forms hills which are surrounded by low-lying Triassic terrain. On the south-western edge of the main mass there is a smaller intrusion of quartz-mica diorite. The contact between this intrusion and the granodiorite is seen at Kinchley Hill, on the eastern bank of the Swithland Reservoir, where inclusions of diorite in the granodiorite suggest that the diorite is the older rock. At Brazil Wood, an island in the Reservoir, the diorite has metamorphosed slates of probable Charnian age to mica-garnet-hornfels.

The age of the Mountsorrel granodiorite has long been in doubt, for the field evidence is not conclusive; it appears to be later than the main Charnian 'syenites' as its structural features are entirely different; moreover, the associated diorite is intruded into beds which probably belong to the Swithland Slates, the highest known Charnian sediments. The granodiorite is cut by dolerite dykes of probable Carboniferous age at Mountsorrel and elsewhere. On the basis of this evidence Watts (1947, p. 76) suggested that the intrusion might be associated with the Caledonian earth-movements. Recent age-determinations by the potassium/argon method (Miller and Podmore 1961, Meneisy and Miller 1963) also suggest that the granodiorite and the associated intrusions are all of 'Caledonian' age. For the present they are best regarded as post-Tremadoc and pre-Carboniferous.

# 6. Carboniferous

The Carboniferous System is the most extensive of the Palaeozoic systems within Central England. Its major subdivisions are as follows:

|  | Approximate maximum thickness ft | (metres) |
|---|---|---|
| Upper Carboniferous | | |
| Coal Measures (Westphalian) | 9000 | (2743) |
| Millstone Grit Series (Namurian) | over 4000 | (1219) |
| Lower Carboniferous | | |
| Carboniferous Limestone Series (Dinantian; subdivided into Viséan and Tournaisian) | over 1500 | (457) |

This broad classification is basically lithological. The Carboniferous Limestone Series consists largely of limestone and dolomite, with local mudstones and sandstones. A rhythmic succession of mudstones and sandstones characterizes both the Millstone Grit Series and the Coal Measures, but the Coal Measures contain less sandstone than the Millstone Grit Series and include many coal seams.

## Carboniferous Limestone Series

Rocks of this Series are exposed at the surface only in a few small inliers. In the western part of the region they occur at Astbury, near Congleton, at Lilleshall and Little Wenlock in the Coalbrookdale Coalfield, and on Titterstone Clee Hill. To the east they are seen at Breedon Cloud and other small inliers on the north-east margin of the Leicestershire Coalfield. They have been proved by boreholes in the Leicestershire Coalfield and in the northern parts of the South Staffordshire and Warwickshire coalfields, in the north-east Midlands between Charnwood Forest and Newark-on-Trent, and at Kettering and Northampton.

The Carboniferous Limestone Series can be separated into two main facies, dependent on deposition in two distinct environments. These are the rigid blocks or 'massifs', relatively stable areas in which subsidence and sedimentation were slow and often delayed, and the 'basins', areas in which subsidence and sedimentation were more rapid. The 'massif' facies is characterized by well-bedded relatively pure limestones with a fauna dominated by corals and brachiopods, with the local development of so-called 'lagoon phase' sediments comprising porcellanous limestones, oolites and dolomites which probably accumulated in restricted areas of shallow water bordering the land. The 'basin' facies sediments comprise dark impure limestones with a poor fauna of solitary corals, and dark shales and thin limestones with a fauna of goniatites and lamellibranchs. In Central England, the Carboniferous Limestone Series is chiefly of the 'massif' facies, the 'basin' facies being confined to the subsurface 'Widmerpool Gulf' (Falcon and Kent 1960) south of Nottingham and Grantham.

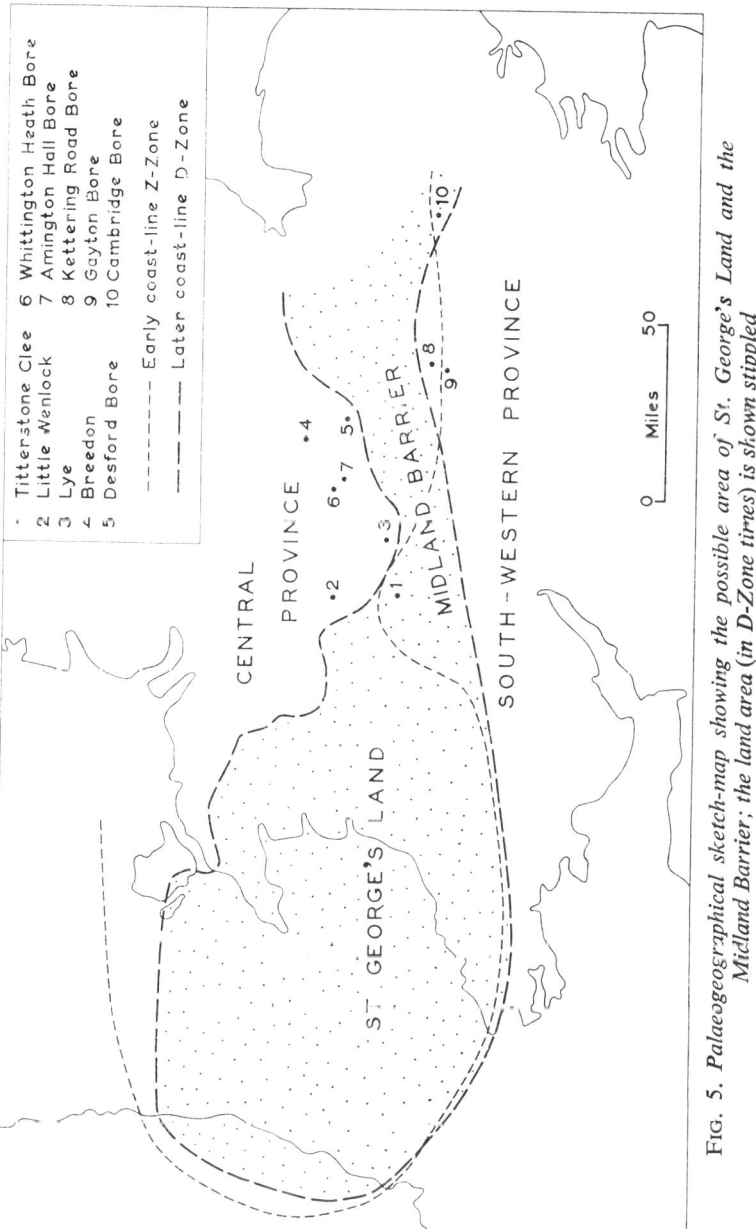

FIG. 5. *Palaeogeographical sketch-map showing the possible area of St. George's Land and the Midland Barrier; the land area (in D-Zone times) is shown stippled*

The dominant palaeogeographical features in Central England during Carboniferous Limestone times were the land areas of St. George's Land and its eastern extension, the Mercian Highlands, which separated the two main areas of deposition the Central and South-Western Provinces (Fig. 5). The Period began with the northward transgression of the sea in the South-Western Province over the Upper Old Red Sandstone delta-plain. During the Tournaisian the sea is known to have extended at least as far north as Titterstone Clee Hill and eastwards to Gayton and Cambridge. In the Viséan there was a marked transgression of the sea into the Central Province, and the Carboniferous rests with strong unconformity on Lower Palaeozoic and Pre-Cambrian rocks. The Viséan thins southward against the Mercian Highlands, being over 1500 ft (457 m) thick in North Derbyshire, whereas at Stockshouse Farm Borehole, near Desford (Fig. 5, No. 5) it is 27 ft 6 in (8·3 m) thick and at Whittington Heath Borehole, near Lichfield (Fig. 5, No. 6) only 2 ft 6 in (0·8 m) thick. In the South-Western Province the sea appears to have retreated from the northern fringe, and the only occurrence of deposits of Viséan age in that part of Central England is in the Kettering Road Borehole, Northampton. The possible shorelines of the Carboniferous Limestone sea in Tournaisian and Viséan times are shown in Fig. 5. It must be emphasized that these shorelines are based mainly on negative evidence (*viz.* the absence of deposits) and should be regarded as largely conjectural.

The Carboniferous Limestone Series was subdivided by Vaughan (1905) into a series of zones based on coral-brachiopod assemblages. Beyond the type area of the Avon Gorge, Bristol, this classification must be applied with caution, as corals and brachiopods are not such reliable zonal indicators as are the more rapidly evolving forms such as goniatites.

The Tournaisian is poorly represented in Central England, but the Viséan is more widely developed and three zonal divisions are recognized:

Viséan
- *Dibunophyllum* Zone (D) (Subzones $D_1$, $D_2$)
- *Seminula* Zone ($S_2$)
- Upper *Caninia* Zone ($C_2S_1$)

Of these, the *Seminula* Zone and the *Dibunophyllum* Zone correspond fairly well with those of the South-West Province. In the upper part of the succession in the basin facies of parts of northern England and the North Midlands, the coral-brachiopod zones cannot be applied, and a classification based on goniatites and lamellibranchs is employed in which the *Beyrichoceras* Stage (B) is equivalent to the $S_2$–$D_1$ Zones, the Lower *Posidonia* Stage ($P_1$) is equivalent to the $D_2$ Subzone, and the Upper *Posidonia* Stage ($P_2$) equates with the subzone formerly termed '$D_3$' in the coral-brachiopod classification.

**Tournaisian**

Tournaisian rocks with a maximum thickness of about 150 ft (46 m) crop out on the northern and southern sides of Titterstone Clee Hill. The lowest beds comprise shales and buff to grey limestones with a basal conglomerate. Their fauna includes *Pugilis vaughani* and *Unispirifer* cf. *tornacensis*. They are succeeded by massive crinoidal limestones with some beds of light grey oolite yielding '*Camarotoechia*' *mitcheldeanensis, Syringothyris cuspidata* and species of the corals '*Zaphrentis*' and *Michelinia*. Fish remains are common and include teeth of *Orodus* and spines of *Ctenacanthus*. The limestones are overlain unconformably by the Cornbrook Sandstone (pp. 37-8).

In the East Midlands, a borehole at Gayton (Fig. 5, No. 9), five miles (8 km) south-west of Northampton, proved red and grey shales with early Carboniferous fossils, and white limestones, resting on coarse red sandstones and marls of Old Red Sandstone type. Carboniferous Limestone with a Tournaisian fauna was proved in a borehole at Cambridge (Fig. 5, No. 10).

**Viséan**

The most northerly outcrop of the Carboniferous Limestone Series within Central England is at Astbury, south of Congleton, where a small inlier occurs in the core of an anticline on the eastern side of the Red Rock Fault. The lowest beds exposed are massive-bedded grey limestones which contain brachiopods and corals indicative of the $D_1$ Subzone. These limestones are overlain successively by shales and thin limestones, agglomerates and tuffs, and further shales with two thin coal seams and a fauna including the cephalopod *Merocanites* and many brachiopods, corals and polyzoa. The highest beds, about 300 ft (91 m) above the basal limestone and probably of $P_2$ age, include sandstones which are overlain by shales of the Millstone Grit Series.

In the Coalbrookdale Coalfield area the Carboniferous Limestone occurs in a small faulted outcrop at Lilleshall, and in a narrow strip, broken by several faults, along the western side of the coalfield north of Ironbridge. It is not known if the Limestone is continuous beneath the Coal Measures between these two outcrops. At Lilleshall, the Limestone, about 280 ft (85 m) thick, is believed to rest unconformably on Cambrian sandstone and is unconformably overlain by Coal Measures. The lowest beds are grey sandstones and limestones, succeeded by grey and red shales and thin limestones with a fauna including the brachiopods *Avonia, Dielasma* and *Productus garwoodi*, the polyzoan *Fenestella* and the coral '*Zaphrentis*'. These beds are overlain successively by nodular limestones, red calcareous sandstone and red and black limestones and shales. The nodular limestones contain a coral fauna indicative of the $D_1$ Subzone which includes *Dibunophyllum bourtonense, Lithostrotion junceum* and *Palaeosmilia murchisoni*. In the western outcrop the lowest bed is the Lydebrook Sandstone, about 70 ft (21 m) thick, which comprises fine to coarse-grained sandstones and conglomerates with subordinate sandy shales. It contains a D Zone fauna including *Gigantoproductus* cf. *maximus*. The Sandstone is succeeded by a creamy white nodular limestone, 2 to 6 ft (0·6 to 1·8 m) thick, followed by 100 ft (30 m) of microporphyritic olivine-basalt. The highest beds, about 50 ft (15 m) thick, consist of impure limestone with bands of sandstone and shale with a $D_2$ fauna including the corals *Lithostrotion junceum* and *Lonsdalcia floriformis* and the brachiopod *Gigantoproductus latissimus*. Pocock (1926) showed that the basalt maintains a constant horizon; he also pointed out that the absence of contact-alteration in the overlying limestone, the occurrence locally of an overlying bed of red bole formed by contemporaneous weathering, and the slaggy upper surface of the rock show it to be a lava-flow. In the Doseley quarries well-defined columnar jointing is developed. North of the coalfield, Carboniferous Limestone has been proved in two boreholes to the south-west of Market Drayton.

In the South Staffordshire Coalfield, Carboniferous Limestone (D Zone) is present beneath the Coal Measures at Fair Oak Colliery (Cannock Chase). Fragments of Lower Carboniferous rocks occur in the basement bed of the Coal Measures at Hilton Main Colliery, near Wolverhampton, and at Wollescote and Lye near Stourbridge; the size of the fragments suggests a local derivation. North of the coalfield, Carboniferous Limestone was recorded, below Bunter Pebble Beds, from a borehole at Chartley, north-east of Stafford, and it forms a small inlier at Snelston Common, four miles (6·4 km) south-south-west of Ashbourne.

A small area of concealed Carboniferous Limestone occurs between Lichfield and the northern end of the Warwickshire Coalfield. A boring at Whittington Heath, near Lichfield, proved 2 ft 6 in (0·8 m) of Carboniferous Limestone of Viséan age at −3670 ft (−1119 m) O.D., unconformably overlain by Millstone Grit and overlying Upper Old Red Sandstone Beds. Carboniferous Limestone of the D Zone with a fauna including *Lithostrotion junceum* and species of *Gigantoproductus* was reached below Millstone Grit in borings at Amington Hall and Statfold at the extreme northern end of the Warwickshire Coalfield.

In the Leicestershire Coalfield, a boring at Ellistown penetrated 86 ft (26 m) into the Carboniferous Limestone. About 5 miles (8 km) to the south at the Stockshouse Farm Borehole, near Desford, the Limestone has thinned to 27 ft 6 in (8 m).

Several small inliers of Carboniferous Limestone occur to the north-east of Ashby-de-la-Zouch. An eastern series comprises the outcrops of Breedon on the Hill, Breedon Cloud, Barrow Hill, Osgathorpe and Grace Dieu; and a western series those of Ticknall, Calke and Dimminsdale. The western outcrops consist mainly of grey limestones and shales which are fairly fossiliferous. At Ticknall the lower beds include crinoidal and foraminiferal limestones with *Dibunophyllum bipartitum, Zaphrentites enniskilleni* and *Gigantoproductus giganteus* succeeded by thinly bedded limestones and shales with abundant Productids. These lower beds, probably of $D_2$ age, are successively overlain by dolomite and by shales which are succeeded by the Millstone Grit. In the easterly inliers poorly fossiliferous grey and yellow dolomites predominate. About 600 ft (183 m) of beds are seen in the quarry at Breedon Cloud. The lowest dolomites exposed, of $C_2$ age, contain moulds of the Productid *Levitusia* [*Plicatifera*] *humerosa*. They are overlain by purple-stained, pebbly, dolomitic limestones with a $C_2S_1$ fauna of corals and brachiopods. Higher beds include a band of dolomite breccia with *Davidsonina* [*Cyrtina*] *septosa* and corals indicating a $D_1$ age which is succeeded by further dolomites and limestones with goniatites and brachiopods of the $B_2$ Zone.

In the Derby–Nottingham–Melton Mowbray area (Widmerpool Gulf) recent exploratory drilling for oil and gas has revealed an east–west subsurface 'gulf' of Lower Carboniferous rocks of the basin facies (Falcon and Kent 1960). The gulf appears to extend westwards to link up with the

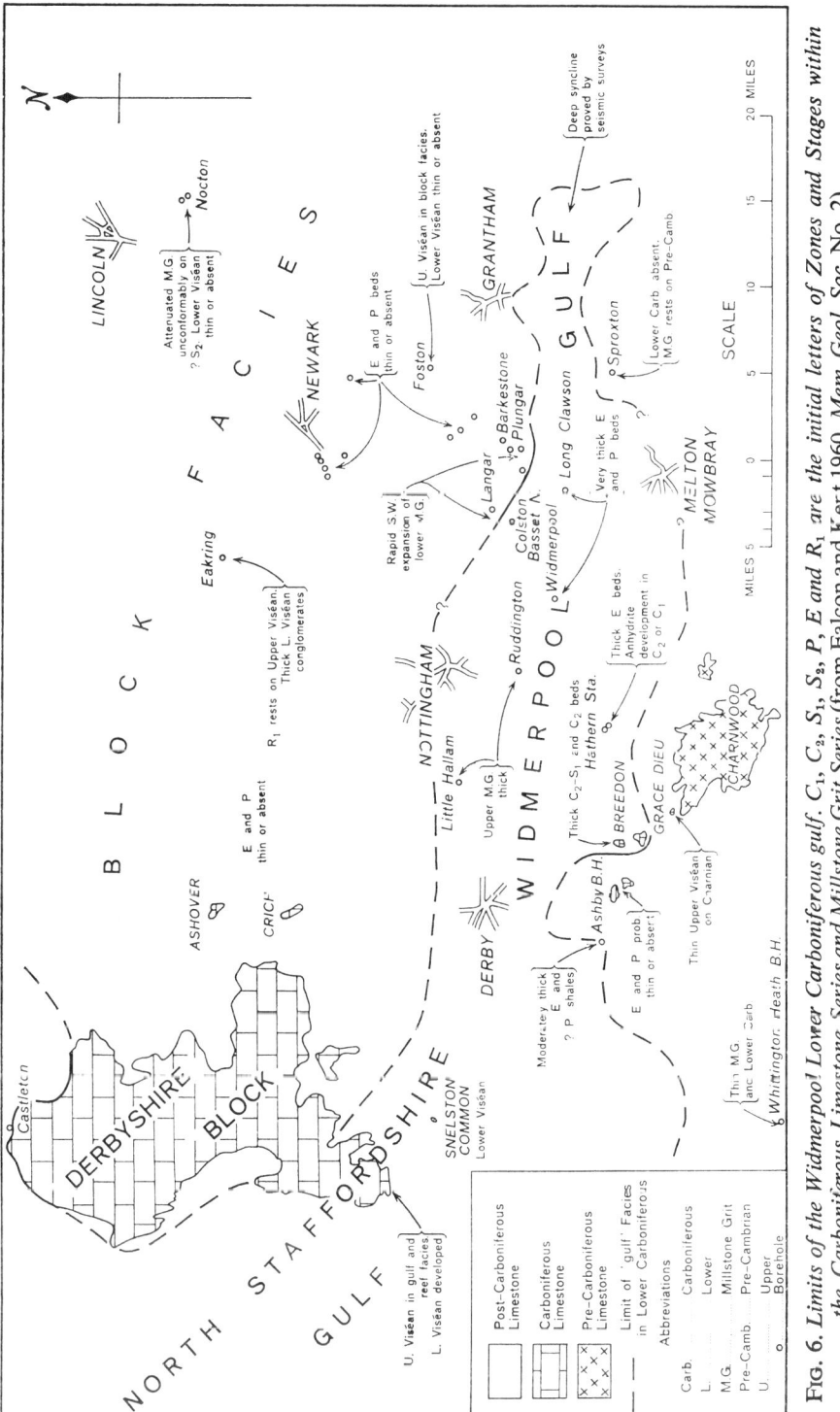

Fig. 6. Limits of the Widmerpool Lower Carboniferous gulf. $C_1, C_2, S_1, S_2, P, E$ and $R_1$ are the initial letters of Zones and Stages within the Carboniferous Limestone Series and Millstone Grit Series (from Falcon and Kent 1960, Mem. Geol. Soc. No. 2)

North Staffordshire area, but its eastward extent is at present uncertain (Fig. 6). The Widmerpool No. 1 Borehole, in the centre of the gulf, penetrated 1454 ft (443 m) into the Carboniferous Limestone Series below a thick sequence of Millstone Grit. The strata, of $P_2$ age in the upper part, comprise mudstones, shales and limestones. Similar beds were present in the Long Clawson Borehole some 6 miles (9·7 km) to the east. The Hathern No. 1 Borehole, 3 miles north-north-west of Loughborough, has revealed the only known occurrence of evaporite deposits within the Carboniferous of Central England. Below the Millstone Grit the borehole passed through 491 ft (150 m) of limestone (probably of $C_2$ age in the lower part) into an alternation of anhydrite, shale and limestone which was proved to an incomplete thickness of 391 ft (119 m). North of the gulf, borings have proved a Carboniferous Limestone sequence of normal massif facies over a wide area east of the Pennines. To the south, the limestone thins rapidly against the Mercian Highlands (p. 30).

## Millstone Grit Series

Lower Carboniferous times closed with the elevation and erosion of the deposits at the margins of the Central Province sea, with the result that over much of this region there is an unconformity between the Millstone Grit Series and the Carboniferous Limestone Series. The Millstone Grit Series was laid down on a series of deltas on the northern side of St. George's Land and the Mercian Highlands. The region of maximum accumulation lay to the north of what is now Central England.

In the southern Pennines the earliest Namurian sediments are predominantly marine; later deposits show evidence of rhythmic sedimentation oscillating between marine and deltaic conditions. Thick beds of grit occur at some horizons. Each grit may be due to a separate period of uplift of the adjacent landmass or to climatic changes leading to greater erosion and a consequent increase in the amount and grain size of the detritus reaching the sea. The grit beds are made up of coarse sandstones with conglomerates. The high feldspar content of the northern sandstones may indicate derivation from a landmass composed largely of igneous and metamorphic rocks. In North Staffordshire the older sandstones are non-feldspathic and may have been derived from a different source. Subsidence at times failed to keep pace with the deposition of the coarse sediments; the deltas were then built up to water-level, and exposed as sandy flats which became colonized by swamp vegetation. This led to the formation of the fireclays and thin coals which overlie some of the grits. Each grit phase was terminated by renewed subsidence, so that the grits became buried beneath fine muds.

The waters in which the bulk of the Millstone Grit shales were accumulated were not normally favourable to marine life, but periodically goniatites (Fig. 7), lamellibranchs, horny and calcareous brachiopods, and fish proliferated. Their remains are found in the 'Marine Bands' which occur at intervals throughout the Series. Successive marine bands are characterized by different species of goniatites which provide a reliable means of correlation of the strata as a whole (p. 36). Drifted plant remains are locally abundant in both grits and shales. The Lower Carboniferous

## Carboniferous

flora persists into the lowest beds of the Series, but it is replaced by a flora of Upper Carboniferous type before the end of the Upper *Eumorphoceras* Stage.

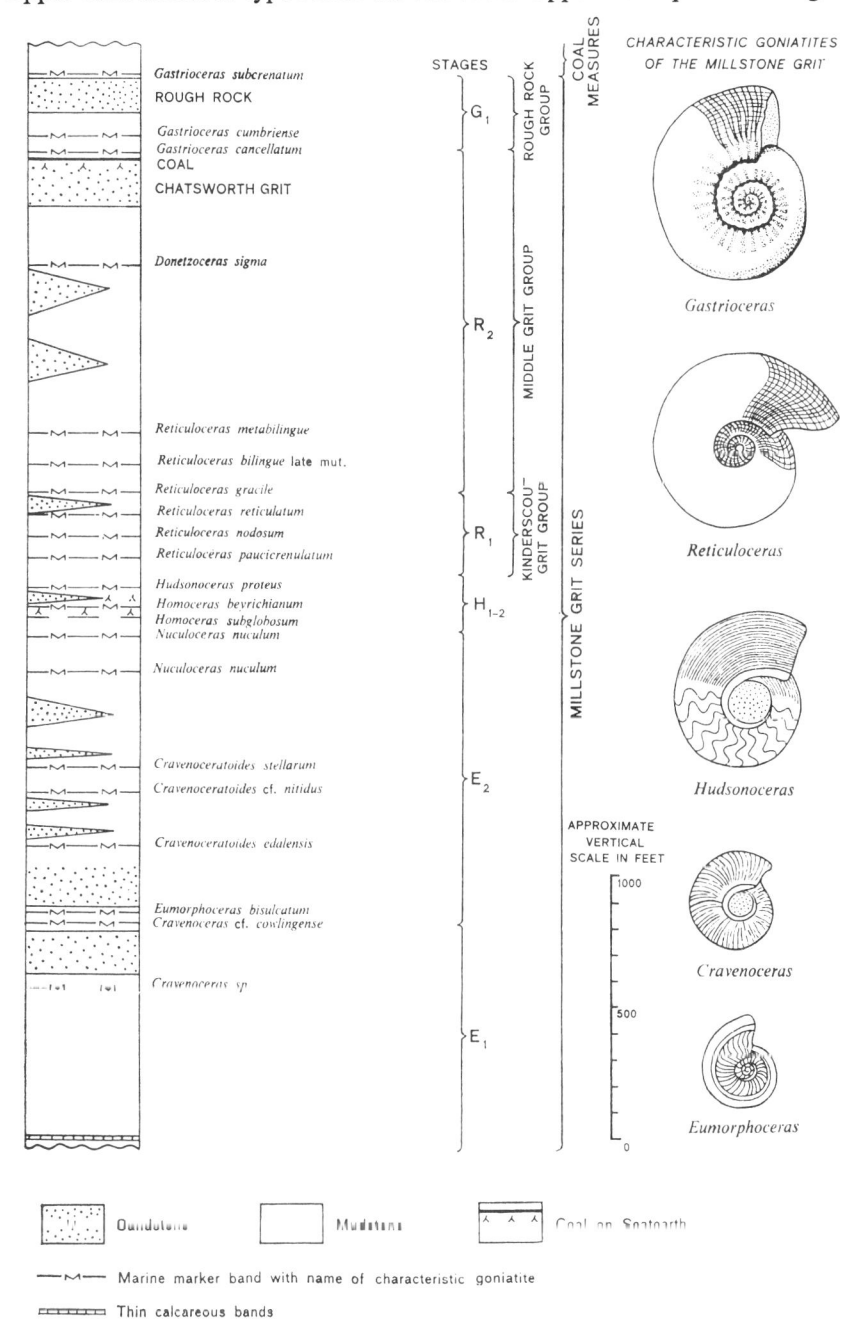

FIG. 7. *The succession and goniatite stages of the Millstone Grit Series of North Staffordshire*

## North Staffordshire

The Millstone Grit Series crops out on the flanking anticlines and associated synclines marginal to the large south-south-westerly plunging syncline of the Potteries Coalfield. A further outcrop forms a rim round the northern side of the Cheadle Coalfield. The Series varies in thickness from over 4000 ft to 2300 ft (1219–701 m) owing to marked thinning to the west and south from near the northern apex of the Potteries Coalfield. The main sandstones, resistant to weathering, form strong escarpments. Valleys have been carved in the softer mudstones, which also form the more gently rising ground below the sandstone escarpments. Although the mudstones of the Series are actually of greater thickness than the sandstones, they are poorly exposed and their outcrops are commonly obscured by downwash from the overlying sandstone escarpments, and by glacial drift.

The Millstone Grit succession in North Staffordshire, shown diagrammatically in Fig. 7, consists predominantly of mudstone. Sandstones are chiefly fine-grained except in the upper part of the sequence which includes two or more coarse-grained members. Unlike the main part of the succession, which may have been laid down exclusively under marine conditions (except for local seatearths), these coarse sandstones were deposited rapidly as fluvial sands and are separated by fossiliferous shales denoting periodic marine incursions.

In this structurally complex area, detailed assessment of variations in the sequence would be difficult were it not for the prevalence of a wide range of goniatites which occur abundantly in some of the darker bands of mudstone, and are used in zoning the strata, as shown in the following table:

STAGES, GROUPS AND ZONES OF THE MILLSTONE GRIT SERIES
(in downward succession)

| | Approximate maximum thickness in Central England |  |
|---|---|---|
| | ft | (metres) |
| STAGE $G_1$ (Lower *Gastrioceras*): ROUGH ROCK GROUP Zones (from above downwards): *G. cumbriense, G. cancellatum* | 300 | (91) |
| STAGE $R_2$ (Upper *Reticuloceras*): MIDDLE GRIT GROUP Zones: *R. superbilingue, R. bilingue, R. gracile* | 1750 | (533) |
| STAGE $R_1$ (Lower *Reticuloceras*): KINDERSCOUT GRIT GROUP Zones: *R. reticulatum, R. nodosum, R. circumplicatile* | 325 | (99) |
| STAGE $H_2$ (Upper *Homoceras*): Zones: *Homoceratoides prereticulatus, Homoceras undulatum, Hudsonoceras proteus* STAGE $H_1$ (Lower *Homoceras*): Zones: *Homoceras beyrichianum, H. subglobosum* | 275 | (84) |
| STAGE $E_2$ (Upper *Eumorphoceras*): Zones: *Nuculoceras nuculum, Cravenoceratoides nitidus, Eumorphoceras bisulcatum* | 1400 | (427) |
| STAGE $E_1$ (Lower *Eumorphoceras*): Zones: *Cravenoceras malhamense, Eumorphoceras pseudobilingue, Cravenoceras leion* | 700 | (213) |

*Stage $E_1$.* Beds of $E_1$ age crop out in the cores of the anticlines flanking the Potteries Coalfield. They are mudstones with scattered lamellibranch bands, overlain by sandstones, totalling over 700 ft (213 m) in thickness.

*Stage $E_2$.* These beds are about 1400 ft (427 m) thick to the east of Biddulph but thin to less than 750 ft (229 m) to the south and west. Basal mudstones with sporadic marine horizons rich in the lamellibranch *Posidonia corrugata* are overlain by sandy formations which form prominent topographic features; these beds are about 500 ft (152 m) thick and consist of fine or medium-grained quartzitic sandstones, with numerous mudstone partings, becoming more silty towards the west. Two marine bands indicative of the lowest zone in $E_2$ occur in a thick shale bed within the sandstones. Higher beds consist largely of unfossiliferous mudstones interspersed with several marine bands with goniatites characteristic of the zones of *Cravenoceratoides nitidus* and *Nuculoceras nuculum*. Several fine or medium-grained quartzitic sandstones occur locally, and because of their compact nature and steep dips they form small, but very distinct features.

*Stages $H_1$ and $H_2$.* The strata comprising these stages have a fairly uniform thickness of about 200 ft (61 m). They include the Stanley Grit, fine-grained quartzitic beds intermittently rich in drifted plants, which appear to be approximately equivalent to a cyclic succession of ganisters and mudstones which crops out on the western slopes of Congleton Edge. In addition to the typical goniatite-lamellibranch faunas, characterized by *Homoceras beyrichianum* and *Hudsonoceras proteus* (Fig. 7), these latter beds contain shales with a benthonic fauna of brachiopods, lamellibranchs and gastropods not recorded at this level elsewhere in the area.

*Stage $R_1$ (Kinderscout Grit Group).* In North Derbyshire this Group is about 1600 ft (488 m) thick and includes the coarse Kinderscout Grit; it thins to the south and near Biddulph it consists of about 300 ft (91 m) of mudstones with a number of marine bands belonging to the three zones in the $R_1$ Stage.

*Stage $R_2$ (Middle Grit Group).* This Group differs from underlying beds by the presence of feldspathic sandstones, of which the thickest and most persistent is the Chatsworth Grit at the top of the Group. This latter sandstone forms conspicuous 'hogsback' features, as at Troughstone Hill northeast of Biddulph. Four miles north-east of Biddulph the Group reaches its maximum thickness of over 1700 ft (518 m) with local thick sandstones in the middle of the sequence. It thins rapidly southwards; east of Stoke-on-Trent it is about 540 ft (137 m) thick and even the persistent Chatsworth Grit appears to die out southwards here.

*Stage $G_1$ (Rough Rock Group).* The mudstones above the Chatsworth Grit are of fairly constant thickness and include marine bands typical of the two zones in $G_1$. These are overlain by the Rough Rock, a persistent member about 110 ft (34 m) thick, coarse-grained and commonly false-bedded. Its hardness varies; thus near Biddulph it forms the tors at Rock End, but within half a mile it is sufficiently soft to be crushed for moulding sand.

**Shropshire**

The Cornbrook Sandstone of Titterstone Clee was originally considered to be of Lower Carboniferous age (Kidston and others 1917). More recently it has been classified by George (1956) on the basis of a structural and lithological comparison with South Wales, as Millstone Grit (Namurian). These authors considered that the Cornbrook Sandstone is overlain unconformably by the 'Productive' Coal Measures. Jones and Owen (1961), however, have concluded that the Cornbrook Sandstone and the Coal Measures are conformable and state that the flora of the Cornbrook Sandstone includes *Neuropteris tenuifolia* and *Lonchopteris rugosa*, indicating that it is of Westphalian B (Middle Coal Measures) age. These difficulties suggest that arenaceous deposits of more than one age may have been grouped with the Cornbrook Sandstone.

The Cornbrook Sandstone, up to 700 ft (213 m) in thickness, consists of coarse feldspathic sandstone with pebbly layers and some mudstone partings. The rocks are lithologically similar to the Millstone Grit of Warwickshire and South Derbyshire. The Sandstone is strongly conglomeratic at the base and rests unconformably upon rocks varying in age from Lower Old Red Sandstone to late Tournaisian. Recent palynological work has shown that the lower part of the Cornbrook Sandstone is of Namurian age, while the upper part may be attributed to the Westphalian. The typical Cornbrook Sandstone is overlain unconformably by about 60 ft (18 m) of fine-grained quartzitic sandstone of Coal Measures age (Church Stretton (Sheet 166) Memoir, p. 256).

**Central Midlands and Widmerpool Gulf**

In the central Midlands the Millstone Grit Series crops out in the northern part of the Leicestershire and Warwickshire coalfields and has also been recorded from several boreholes. In comparison with North Staffordshire, the sequence is greatly attenuated, and the Series dies out southwards against the old landmass of the Mercian Highlands.

At the northern end of the South Staffordshire Coalfield, a borehole at Rugeley proved a possible 76 ft (23 m) of Millstone Grit Series with goniatites including *Gastrioceras cancellatum* and *G. crencellatum* in the upper part. The Whittington Heath Borehole (Fig. 5, No. 6) penetrated 196 ft (60 m) of beds assigned to the Series, although only the $R_2$ and $G_1$ stages were definitely recognized. The base of the Series was marked by a thin breccia bed. A comparable thickness of Millstone Grit was proved in two boreholes near Tamworth, at the northern end of the Warwickshire Coalfield, where its base was again marked by a thin breccia bed. 74 ft (22 m) of Millstone Grit were proved in a borehole at Merevale; it crops out nearby and also on the western side of the coalfield at Dosthill. No Millstone Grit has been recorded from the southern parts of the South Staffordshire and Warwickshire coalfields. Recent work on the northern side of the Leicestershire Coalfield, near Melbourne, has led to the recognition of attenuated representatives of the $G_1$ and $R_2$ stages. The lowest beds of the Series have not yet yielded diagnostic fossils, but it is possible that the lower stages are absent. Several boreholes within the Leicestershire Coalfield have proved attenuated sequences of the Millstone Grit Series. At the Stockshouse Farm Borehole, near Desford (Fig. 5, No. 5) the Series is apparently only 12 ft (3·7 m) thick, though the absence of diagnostic fossils makes it difficult to distinguish between Lower Coal Measures and Millstone Grit.

A thick Millstone Grit sequence is developed in the Widmerpool Gulf (Fig. 6). The Widmerpool No. 1 Borehole, in the centre of the Gulf, penetrated about 2600 ft (792 m) of Millstone Grit Series, with Carboniferous Limestone Series below. The strata were mainly grits and sandstones in the upper part, with shales below with an $E_2$ fauna near the base. To the east, at Long Clawson, the Series has thinned to about 1500 ft (457 m) and the sandstones are less well developed. In both boreholes there are intrusions of dolerite in the lower part of the Series. South of the Gulf, at Sproxton, the Series has thinned to 245 ft (75 m) and appears to rest unconformably on Pre-Cambrian beds. On the northern side of the Gulf, the normal Pennine succession of the Millstone Grit extends eastward as far as Eakring and Plungar, but there is a rapid thinning of the grits farther to the east.

## Coal Measures

The Coal Measures have been classified into three major divisions—Lower, Middle and Upper (Stubblefield and Trotter 1957) (see p. 44). In general terms, within Central England the Lower and Middle Coal Measures consist mainly of grey shales with coal seams, while in the Upper Coal Measures red marls and sandstones predominate. There are many exceptions to these broad lithological divisions.

The Lower and Middle Coal Measures were laid down on the deltas initiated during Millstone Grit times. North of the Midland Barrier (Mercian Highlands) deltas appear to have covered almost the whole of the north Midlands, northern England and southern Scotland. The main source of sediment was probably a North Atlantic Continent, with only a little detritus being derived from St. George's Land and the Midland Barrier. From time to time the supply of sediment exceeded the rate of subsidence, and extensive sub-aerial delta-flats were formed. Forests developed on the flats, and the resulting plant debris was eventually buried under sediment (and converted to coal seams) when further subsidence took place. Occasionally the delta-flats were invaded by the sea, but although such marine incursions extended over wide areas they were short-lived.

Although the bulk of the Lower and Middle Coal Measures sediments are of shallow water or terrestrial origin, subsidence of the pre-Carboniferous floor allowed a great thickness of sediment to accumulate and in the area of maximum subsidence, to the north of Central England, more than 5000 ft (1524 m) of beds were laid down. During the subsidence the area of deposition spread southwards, and deltas gradually encroached on the Midland Barrier. Consequently the lowest Coal Measures are present only in the northern part of the region, where they rest conformably on the Millstone Grit Series; higher beds successively overlap the lower beds southwards, and in some areas rest directly on pre-Carboniferous rocks.

Variations in the original thickness of the Lower and Middle Coal Measures were partly produced by differential subsidence; some areas subsided rapidly while others remained more stable or even showed a tendency to rise. This can be seen by considering the Thick Coal and its

40  Central England

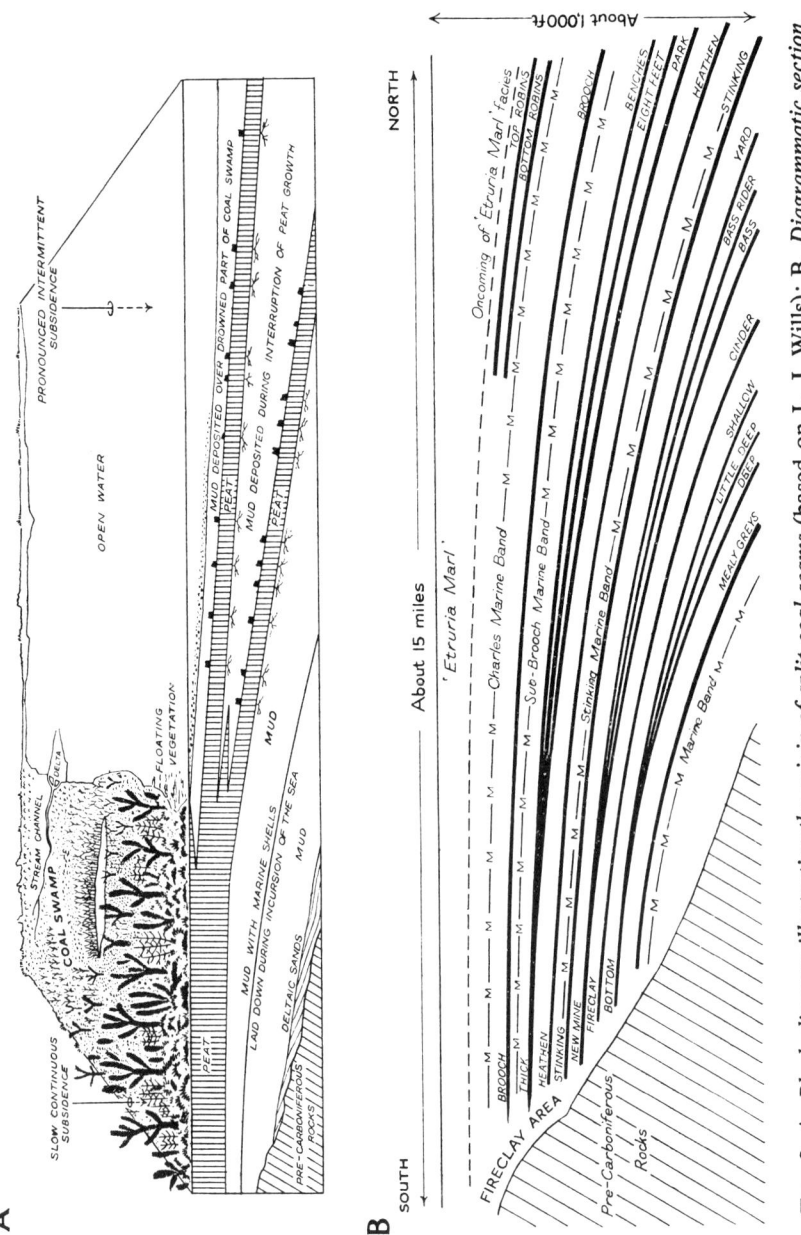

FIG. 8. A. Block-diagram illustrating the origin of split coal seams (based on L. J. Wills); B. Diagrammatic section showing the northward splitting of coal seams in the South Staffordshire Coalfield, and the correlation of seams on the evidence provided by marine bands (based on Jukes, with modifications by G. H. Mitchell and Sir James Stubblefield)

equivalents in the South Staffordshire Coalfield. In the Dudley area, where the Thick Coal is about 30–35 ft (9–11 m) thick, there was very slow subsidence with continuous accumulation of plant debris over a long period. Farther north, in Cannock Chase, subsidence was greater and the formation of the beds of plant debris was interrupted by the deposition of layers of sandy or muddy sediment. Consequently, when traced north, the Thick Coal splits into several distinct seams (Fig. 8). In the Cannock Chase area there are about 170 ft (52 m) of deposits which are equivalent to the 30 ft (9 m) seam at Dudley.

The conditions prevailing during the deposition of the greater part of the Upper Coal Measures differed from those under which the Lower and Middle Coal Measures were laid down. Due to the Carboniferous earth-movements, mountain ranges were rising in areas bounding the basin of deposition. This led to an extension of more continental conditions of deposition, the formation of 'red beds' and the disappearance of the delta swamps. The earliest signs of this change in sedimentary conditions are seen in the Wyre Forest Coalfield, where beds of 'Etruria Marl' facies (red and mottled marls and coarse sandstones) occur within the Lower and Middle Coal Measures. On the margins of the basin of deposition, in Wyre Forest and Coalbrookdale, there is a marked unconformity between the Upper Coal Measures and lower beds ('Symon Fault', Fig. 11). The transition from deltaic to 'red bed' facies was not simultaneous over the whole of Central England and in North Staffordshire deltaic conditions persisted well into Upper Coal Measures times. There was a temporary return to deltaic conditions in the Newcastle-under-Lyme Group and Halesowen Beds, but the coals in these groups are mainly thin and sulphurous.

In the Central England Region the Middle Coal Measures tend to be thicker in the north than in the south. Owing to the erosion of the Upper Coal Measures during Permo-Triassic times, the original thickness of the Upper Coal Measures is unknown, but in North Staffordshire they attain a thickness of nearly 5000 ft (1524 m). The various sub-divisions thin out southwards against the Midland Barrier.

The effects of rhythmic sedimentation, characteristic of the Millstone Grit Series, are also seen in the rocks of the Lower and Middle Coal Measures, though there are considerable lithological differences. In the Lower and Middle Coal Measures, coal seams are thicker and more common, marine mudstones are infrequent, sandstones are thinner, finer grained and less persistent. The 'complete' rhythm consists of seatearth, overlain successively by coal, marine mudstone, non-marine mudstone, siltstone, sandstone and then seatearth again. It is unusual for all the components of the rhythm to be present, and there is considerable variation in the sequence and also rapid lateral variation, especially in the sandstones.

Coal seams comprise only about 2 per cent of the thickness of the Lower and Middle Coal Measures, the main rock types being mudstone and sandstone. The coals are mainly classed as bituminous, with less common seams of cannel. They vary in thickness from thin films to about 36 ft (11 m) but only exceptionally exceed 10 ft (3 m). The mudstones or shales are usually grey or black, and sometimes contain much carbonaceous debris. Shell bands, either marine or, more commonly, non-marine, occur in mudstones above many of the coals. The sandstones are usually fine-grained and well cemented, in contrast to the coarse-grained sandstones and grits of the Millstone Grit Series. They are generally lenticular, and pass laterally into siltstones and mudstones. Cross-bedding is common, and the sandstones sometimes fill channels which are incised in the underlying beds. Where the channels have cut through a coal seam, they are termed 'wash-outs' (Fig. 9). Coal seams are almost always underlain by a rootlet bed, or seatearth. The lithology of these beds can vary from mudstone ('fireclay') to sandstone, depending on the type of sediment upon which the coal-forming vegetation grew. Bands and nodules of sideritic clay-ironstone are common in the mudstones and seatearths. They probably formed by the segregation of ferrous carbonate formed under reducing conditions in muds saturated with iron-rich water and containing decaying vegetation. Blackband ironstones, which contain a high proportion of carbonaceous matter, are uncommon in the Lower and Middle Coal Measures, but occur abundantly in the lower part of the Upper Coal Measures of North Staffordshire (p. 53).

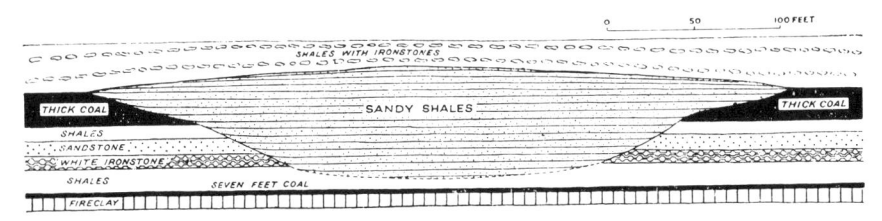

FIG. 9. *Diagram of a 'wash-out' in the Thick Coal of Warwickshire: Victoria Colliery, Hawkesbury*

Although widespread, rhythmical sedimentation is less obvious in the 'red beds' facies of the Upper Coal Measures. The chief sedimentary types, reflecting the change to more arid conditions of deposition, are red mudstones and siltstones ('marls') and purple sandstones. Lenticular beds of green or white grit and conglomerate ('espleys') occur in parts of the Etruria Marl facies. The Keele Group contains occasional thin beds of porcellanous limestone which resemble the cornstones of the Old Red Sandstone and commonly contain the worm *Spirorbis*.

Most of the animal remains occur in mudstones overlying coals or seatearths, and they tend to occur in the following upward sequence of phases: dark mudstone or shale with fish remains, occasionally with non-marine shells or '*Estheria*'; marine mudstone with brachiopods and molluscs; mudstone with non-marine lamellibranchs, occasionally with '*Estheria*' in the lower part. These phases are seldom all present above any one coal or seatearth, and sometimes only the non-marine phase, sometimes no fossiliferous phase at all, is found.

The marine mudstones generally contain *Lingula mytilloides*. When this fossil is not accompanied by other marine organisms the mudstone may be described as a '*Lingula* band'. However, the principal marine bands commonly yield a more varied fauna including other brachiopods such as

Plate V

A. Titterstone Clee Hill, Shropshire   A.9512

(*For full explanation see p. viii*)

B. Downton Castle Sandstone, Onibury, Shropshire   A.9528

Plate VI    Characteristic fossil plant remains from the Coal Measures

1. Clubmoss bark: *Sigillaria tessellata* Brongniart; 2. Pithcast; stem of horsetail; *Calamites suckowi* Brongniart. 3. Part of seed-fern leaf; *Neuropteris gigantea* (Sternberg). 4. Clubmoss bark: *Lepidodendron wortheni* Lesquereux. 5. Gymnospermous seeds: *Carpolithus membranaceus* Goeppert. 6. Part of seed fern leaf: *Linopteris münsteri* (Eichwald). 7. Part of seed fern leaf: *Sphenopteris dilatata* Lindley and Hutton. 8. Horsetail leaf: *Annularia microphylla* Sauveur. 9. Clubmoss bark: *Sigillaria decorata* Weiss.

*Orbiculoidea*, Productids and Chonetids; lamellibranchs including *Dunbarella* and *Myalina*; goniatites, especially *Gastrioceras* in the lower beds and *Anthracoceras* in the higher beds; nautiloids, gastropods, foraminifera and conodonts.

The most important non-marine fossils are the lamellibranchs (mussels) such as *Carbonicola, Curvirimula, Anthraconaia, Anthracosia, Anthraconauta, Anthracosphaerium* and *Naiadites*. Other forms include small crustacea such as the ostracods *Carbonita* and *Geisina* and the branchiopods '*Estheria*' and *Leaia* (particularly in the Upper Coal Measures), larger arthropods such as *Belinurus* and *Euproops*, and fish including the forms *Elonichthys* and *Rhadinichthys*.

Apart from roots (*Stigmaria*) in the seatearths, the flora of the Coal Measures consists of debris of land plants which was carried into the area of deposition and occurs in all types of sediments. The plants most commonly found in the Coal Measures (Pl. VI) are the club-mosses (Lycopodiales, e.g. *Lepidodendron, Sigillaria*), horsetails (Equisetales, e.g. *Calamites*) and seed-ferns (plants with fern-like leaves, but bearing seeds, belonging to the Pteridospermae, e.g. *Alethopteris, Neuropteris*). Coal itself is formed from plant-material, and contains large numbers of spores.

Many classifications have been used to divide the Coal Measures within the coalfields of Central England, but it is only within recent years that anything approaching detailed correlation has been possible and even now there are many unsolved problems (Plate VII). Three groups of fossils—plants, non-marine lamellibranchs and marine faunas—have been used as a basis for dividing the Coal Measures. To some extent the three schemes are complementary, and their inter-relationships are shown in Table 1.

On the basis of fossil plants, Kidston (1894, 1905) proposed a four-fold division of the Coal Measures. As modified by Crookall (1931, 1955) these divisions can be applied with some success over most of Britain. The plant classifications suffer from vague divisional boundaries, although this difficulty has been partly overcome by the use of three marine bands to delimit the lower two divisions of the Westphalian, as recommended by the 1927 Heerlen Congress (Table 1) (Jongmans 1928). Spores have proved useful for correlation purposes, particularly for tracing individual coal seams within a coalfield.

The use of non-marine lamellibranchs (mussels) for correlation, advocated by Hind in the latter part of the last century, fell into disfavour for a considerable time as the results obtained appeared to conflict with classifications based on plants. Trueman and other workers showed that a zonal system based on these lamellibranchs could have practical value, and they are now used to a greater extent than plants. The non-marine lamellibranch zones also suffer from some vagueness in their limits, but this is partly overcome by the use of arbitrary boundaries (coal seams, marine bands) to delimit them. The development of the zonal classification has been summarized by Trueman and Weir (1946, pp. i-xxxii).

Marine faunas occur at a number of horizons within the Lower and Middle Coal Measures. They have been used for many years for correlation within individual coalfields and it is now known that the chief marine bands are so widespread that they can be used for correlation between coalfields

TABLE 1  Classification of the Coal Measures

| MAJOR DIVISIONS | PLANT CLASSIFICATIONS | | NON-MARINE | IMPORTANT |
|---|---|---|---|---|
| Stubblefield and Trotter 1957 | Heerlen Congress 1927 | Kidston and Crookall | LAMELLIBRANCH ZONES | MARINE BANDS |
| UPPER COAL MEASURES | Westphalian D | Radstock Group of Radstockian | *Anthraconaia prolifera* and *Anthraconauta tenuis* | |
| —M——M— | | Farrington Group of Radstockian | | |
| | | Staffordian | *Anthraconauta phillipsii* | *Anthracoceras cambriense* (Bay) |
| —M——M— | Westphalian C | —M——M— | —M——M— | |
| MIDDLE COAL MEASURES | —M——M— | Yorkian | Upper *Anthracosia similis* and *Anthraconaia pulchra* | *Anthracoceras hindi* or *A. aegiranum* (Gin Mine=Charles= Chance Pennystone =Eymore Farm= Nuneaton=Overseal) |
| | Westphalian B | | —M——M— Lower *A. similis* and *A. pulchra* | |
| —M——M— | —M——M— | | Upper *Anthraconaia modiolaris* | *Anthracoceras vanderbeckei* (7-Ft Banbury= Stinking=Penny- stone=Seven Feet= Molyneux=Bag- worth) |
| LOWER COAL MEASURES | Westphalian A | | —M——M— Lower *A. modiolaris* | |
| | | | *Carbonicola communis* | |
| —M——M— | —M——M— | Lanarkian (=Pre-Yorkian) down to 'plant- break' of Scotland | *Anthraconaia lenisulcata* —M——M— | *Gastrioceras sub- crenatum* |
| MILLSTONE GRIT SERIES | Namurian | | | |

(From Edwards and Trotter 1954, p. 46, with additions and amendments) and, in some cases, with coalfields on the Continent. Stubblefield and Trotter (1957) have retained the old three-fold division of the Coal Measures into Lower, Middle and Upper, but have redefined the limits of these major divisions in terms of marine bands. The basal limits of the Lower and Middle Coal Measures are taken at the bases of the marine bands characterized by *Gastrioceras subcrenatum* and *Anthracoceras vanderbeckei* respectively. The latter marine band occurs near the middle of the *Anthraconaia modiolaris* Zone and can easily be recognized even in the absence of the eponymous goniatite. The base of the Upper Coal Measures is taken at the top of the marine band characterized by *Anthracoceras cambriense*. In Central England this marine band has been recognized only in parts of the North and South Staffordshire coalfields, being absent elsewhere due to unconformable relationships between the New Red Sandstone or Upper Coal Measures and the underlying Middle Coal Measures.

The combined use of marine and non-marine faunas has proved the most satisfactory method of correlation within the Lower and Middle Coal Measures.

**Coalfields of Central England**

The Coal Measures originally formed a thick continuous sheet of strata covering much of Central England. This sheet was folded and faulted by the Hercynian earth-movements, dissected by erosion and then buried deeply below Triassic sediments. Subsequent erosion has removed much of the Triassic cover, and parts of the folded and faulted sheet have now become the seven detached coalfields of the region. In some areas of younger rocks which border the visible coalfields, boreholes have proved that the Triassic rests directly on pre-Carboniferous formations, the whole of the Coal Measures having been removed by pre-Triassic erosion. Nevertheless, Coal Measures are preserved beneath the Triassic cover over considerable areas.

The seven coalfields are those of North Staffordshire, Leicestershire and South Derbyshire, Warwickshire, South Staffordshire, Wyre Forest and Coalbrookdale and there are also small areas of Coal Measures on the Clee Hills; all belong to the same (Midland–Pennine) coalfield province. The exposed coalfields occupy an aggregate area of over 600 square miles (1554 km$^2$). The Lower and Middle Coal Measures are the chief productive group; they yield important house, steam, manufacturing and coking coals. The Upper Coal Measures in most areas contain only a few thin, largely unworkable, seams of sulphurous coal.

*The North Staffordshire coalfields* comprise the Potteries Coalfield on the west, and two smaller ones on the east—the Shaffalong and Cheadle coalfields.

The Potteries Coalfield covers a triangular area of just over 100 square miles (259 km$^2$). Its broad structure is that of a syncline plunging to the south-south-west. In this southerly direction the Coal Measures are preserved over the crests of the flanking anticlines which farther north bring up Millstone Grit. Westwards they are thrown down by the Red Rock Fault complex beneath the Triassic of the Cheshire Plain. Eastwards they extend into the long narrow trough of the Shaffalong Coalfield, where the Lower Coal Measures are preserved, before being separated by a sharp

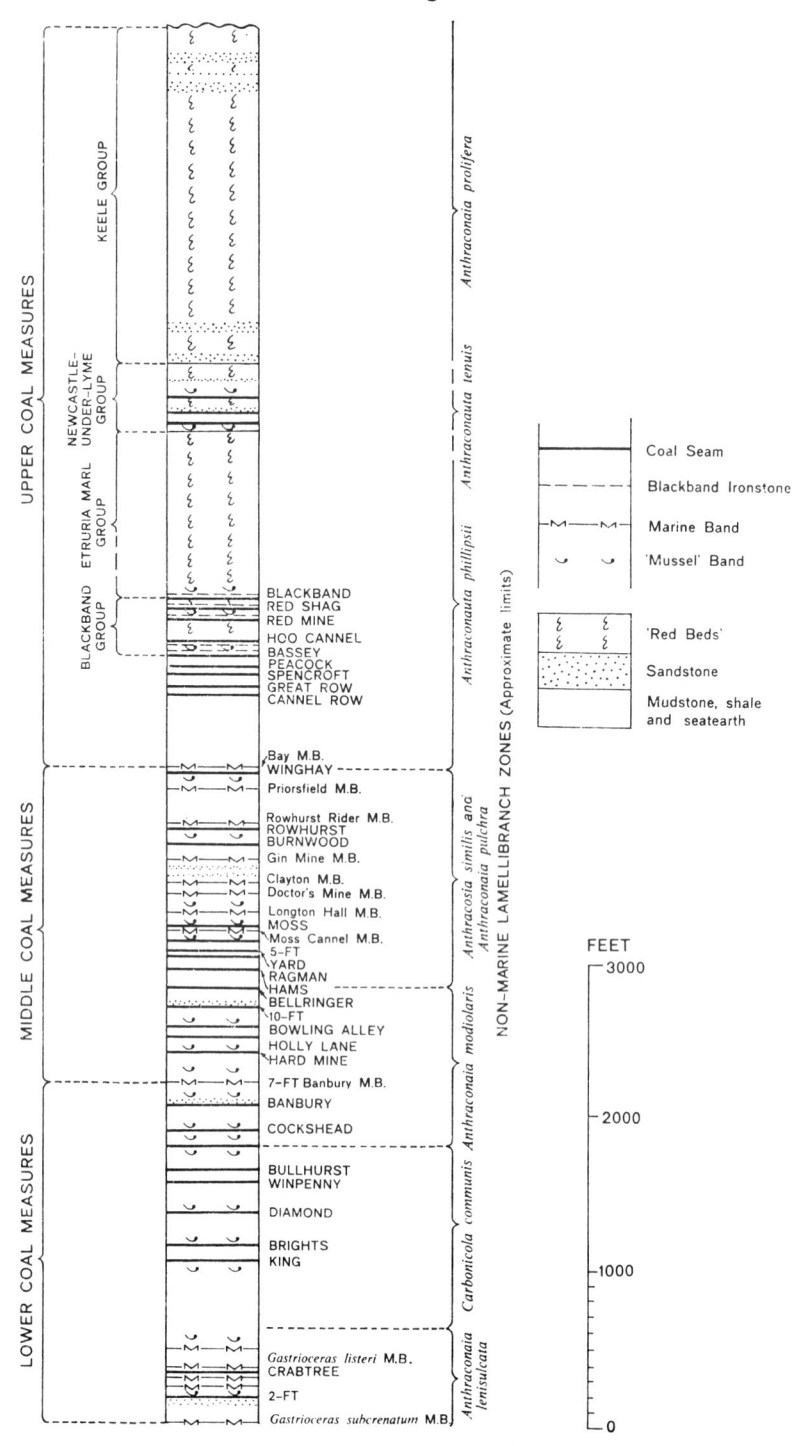

FIG. 10. *Generalized sequence and classification of the Coal Measures of the North Staffordshire (Potteries) Coalfield.* (Based on Maer Borehole (upper 3000 ft) and Wolstanton Colliery Shafts)

faulted anticline from the small Cheadle Basin. To the south Triassic rocks form the base of the triangle. Although no marked unconformity is visible near the axis of the syncline the Bunter Pebble Beds transgress the entire Coal Measures on the eastern limb of the syncline and behave similarly in the Cheadle Coalfield. Strong N.N.W.-S.S.E. faulting cuts the coalfield, the largest individual fracture being the Apedale Fault with a maximum throw of about 2000 ft (610 m). Another set of faults swings from W.-E. to N.W.-S.E. when traced southward.

*The Leicestershire and South Derbyshire coalfields* have a combined area of about 76 square miles (197 km$^2$), all but 24 square miles (62 km$^2$) being concealed beneath a cover of Triassic beds; they are separated by the south-south-easterly plunging Ashby Anticline.

The Leicestershire Coalfield, on the eastern side of the Ashby Anticline, is structurally simple. It consists of the western half of an elongated basin, largely concealed beneath the Triassic, which is truncated on the eastern side by the pre-Triassic Thringstone Fault.

The South Derbyshire Coalfield is bounded to the east by the Boothorpe Fault which lies along the eastern flank of the Ashby Anticline. Although basically synclinal in structure, the coalfield is extensively folded and faulted, with a number of faults with throws of the order of 1000 ft (305 m). The major structures have approximately the same trend as the Ashby Anticline. There is an important concealed extension of the coalfield to the west of the exposed area.

Only Lower and Middle Coal Measures are present in the exposed parts of these coalfields, with the lower (unproductive) beds of the Lower Coal Measures cropping out along the Ashby Anticline. Upper Coal Measures are present beneath the Triassic in the western part of the South Derbyshire Coalfield. Although the two coalfields are in close proximity the successions differ in many respects and correlation between them has proved difficult, especially in the Middle Coal Measures.

*The Warwickshire Coalfield* is roughly oval in outline and covers an area of about 150 square miles (389 km$^2$) extending southwards from Tamworth to Warwick. It is only about 8 miles (13 km) across at its widest part. The Lower and Middle Coal Measures crop out in a narrow belt that swings around the northern end of the coalfield and extends southwards along the eastern margin as far as Bedworth. The greater part of the exposed surface of the coalfield is occupied by barren Upper Coal Measures and Enville Beds. The coalfield consists of a broad synclinal structure with steeply dipping flanks and a gentle southerly plunge. The Lower and Middle Coal Measures rise almost vertically against the Cambrian at Dosthill on the western flank of the syncline, and rise steeply against the Upper Old Red Sandstone and Cambrian to the east between Atherstone and Bedworth. The gentle southward plunge of the fold is interrupted by the Arley Anticline, associated with the Arley Fault (Fig. 17). Much of the coalfield is bounded by faults or erosional fault scarps and steep unconformable contacts suggest that these faults may have been active in Triassic times. Sub-Triassic extensions of the coalfield are present around Tamworth, and between Exhall and Binley.

*The South Staffordshire Coalfield* extends for some twenty-five miles (40 km) between Rugeley in the north and the Lickey Hills in the south, and is bounded to east and west by faults which are nearly ten miles (16 km) apart in the central area. A north-western sub-Triassic extension reaches at least as far north as Sandonbank, three miles north-east of Stafford. To the east of Rugeley a valuable sub-Triassic extension has also been proved, and is being worked by a new pit at Lea Hall, 1 mile (1·6 km) south-east of Rugeley. A full Coal Measures sequence has been proved as far south-east of Rugeley as Whittington Heath, near Lichfield. The Coal Measures rest directly on an irregular floor of Silurian and Downton Series rocks which locally form the steep-sided subsurface 'Silurian banks' (p. 17). Structurally, the coalfield consists of a denuded north–south anticline, separating the Wolverhampton and Birmingham sub-basins, which are now Triassic-covered. Silurian rocks are brought to the surface in the three minor anticlines of Dudley, Sedgley, and Barr, with intervening structurally complex synclinal areas.

*The Coalbrookdale Coalfield* extends southwards from Lilleshall, near Newport to Linley, and covers an area of about 18 square miles. Its north-west margin is formed by the west-south-westerly trending Lilleshall Fault, and a number of strong sub-parallel faults affect the whole area of the coalfield.

The Coal Measures comprise a lower productive group (Lower and Middle Coal Measures) unconformably overlain by Upper Coal Measures. The Lower and Middle Coal Measures were folded on north-easterly axes, and there was considerable erosion before the deposition of the Upper Coal Measures. Consequently the thickness of the Lower and Middle Coal Measures is very variable, with a full sequence preserved only in the synclinal areas (Fig. 11). Beneath this unconformity, known locally as the 'Symon Fault', there is commonly reddening of the Lower and Middle Coal Measures up to a thickness of 60 ft (18 m). Between Linley and the northern end of the Forest of Wyre Coalfield the Upper Coal Measures rest directly on Silurian and Old Red Sandstone strata.

Coalbrookdale is closely connected with the early history of the iron industry. Iron was here first produced on a commercial scale by smelting ironore with coke, and in 1779 the first iron bridge was erected across the Severn at Ironbridge, where it still stands.

*The Forest of Wyre and Clee Hills coalfields.* The Forest of Wyre Coalfield commences near Bridgnorth, and is joined to the Coalbrookdale Coalfield to the north by a narrow outcrop of Upper Coal Measures. Southwards the Coal Measures outcrop broadens and extends along the Severn Valley to the latitude of Bewdley. The Coalfield is joined to the Mamble Coalfield, to the south, which extends to the Abberley Hills. The western margin of the coalfield is defined by the main outcrop of the Old Red Sandstone. The eastern margin is formed by the Enville Fault except at Heightington and Trimpley where inliers of Old Red Sandstone are brought up along the Malvern Axis. The Coal Measures rest unconformably on Old Red Sandstone. The lowest beds are the Kinlet Beds (of Middle and Lower Coal Measures age) which are overlain by Upper Coal Measures (Highley Beds and younger strata).

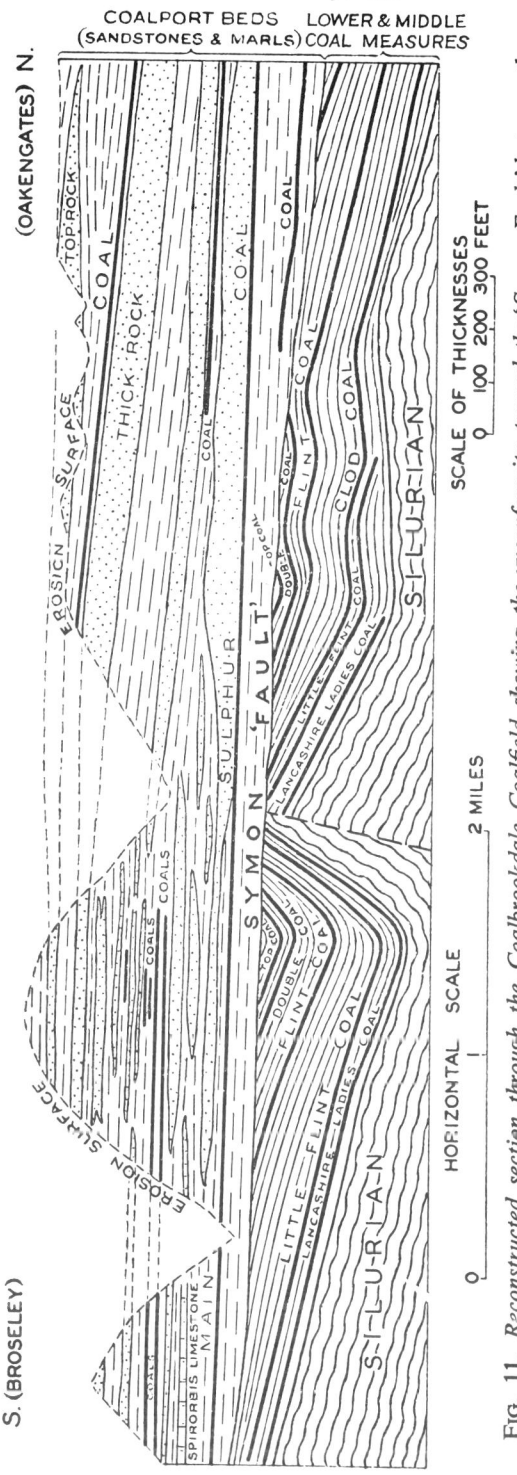

Fig. 11. *Reconstructed section through the Coalbrookdale Coalfield showing the unconformity termed the 'Symon Fault' between the Upper Coal Measures (Coalport Beds) and the Lower and Middle Coal Measures* (after T. H. Whitehead)

Folds subsequent to the deposition of the Coalport Beds have been eliminated by taking the Sulphur Coal as a datum line; in consequence the erosion-surface in the diagram does not represent the present topography

Coal Measures resting unconformably on the Cornbrook Sandstone form the prominent outlier of Titterstone Clee Hill. They comprise a basal sandstone, fine-grained, buff or light yellow in colour which is overlain by shales and sandstones with bands of red clay. Plants indicating a Westphalian B age were recorded from the 'Cornbrook Sandstone' by Jones and Owen (1961) but it is now thought that these may have come from the basal sandstones and shales of the Coal Measures (p. 38). The Productive Coal Measures which overlie the basal Sandstone include five coal seams which have been worked in the past. Several shafts pass through a dolerite sill to reach the highest or Great Seam. Plant remains suggest a Westphalian Cage for the productive measures (Jones and Owen 1961). A small outcrop of Coal Measures, capped by dolerite, occurs on Brown Clee Hill.

**Lower Coal Measures**

Throughout the coalfields of Central England the lower beds of the Lower Coal Measures (*lenisulcata* Zone) contain few workable coals, and they have been separated locally as the 'Lower Unproductive Coal Measures'. In contrast, the higher beds of the Lower Coal Measures contain many of the best workable seams (Plate VII).

In North Staffordshire the Lower Coal Measures comprise some 2000 ft (610 m) of mudstones and siltstones between the *Gastrioceras subcrenatum* Marine Band and the Seven-Feet Banbury Marine Band. Sandstones are infrequent, the thickest and most persistent lying immediately below the Two-Foot seam and close above the Banbury Coal. In the lower three-quarters of the succession coals are numerous but in general are thin. The seams are thicker in the higher beds and include the Bullhurst, Cockshead and Banbury coals which are among the most valuable seams in the Coalfield. Near the base of the Lower Coal Measures there are five marine bands of which the *Gastrioceras listeri* Marine Band, in the roof of the Crabtree Coal, is particularly well developed. Rich assemblages of non-marine lamellibranchs (mussels) occur at several horizons.

The Lower Coal Measures of the South Staffordshire Coalfield thin southwards from about 900 ft (274 m) to about 350 ft (107 m), this thinning being accompanied by a coalescing of some of the seams (Fig. 8B). The chief seams in the northern part of the coalfield, where all the working collieries are now situated, are the Deep and Shallow (which combine to form the Bottom Coal in the south), Cinder, and the Bass and Yard (which combine to form the New Mine Coal in the south). Two marine bands occur in the lower part of the sequence in the northern (Cannock) area including the *G. subcrenatum* Marine Band at the base (known only from the extreme north of the coalfield). Only the higher of the two bands is represented farther south where the Lower Coal Measures rest unconformably on older rocks. The most important mussel band is that above the Cinder Coal. Its fauna includes the large mussel *Carbonicola cristagalli*, and it can be correlated with the bands above the Deep Rider of north Warwickshire, Woodfield of South Derbyshire and Middle Lount of Leicestershire.

In Coalbrookdale the Lower Coal Measures reach a maximum thickness of about 300 ft (91 m). Locally they have been partly or wholly removed by pre-Upper Coal Measures erosion (Fig. 11) and they rest unconformably on Lower Carboniferous or older rocks. The principal worked seams are the Best, Randle and Clod. A marine band occurs in the lower part of the succession.

It is uncertain if any Lower Coal Measures are present in the Forest of Wyre Coalfield. The lowest 300 ft (91 m) of the Coal Measures are of Etruria Marl facies, and no significant fossil horizons have yet been found; they rest unconformably on Old Red Sandstone.

The Lower Coal Measures of Warwickshire thin southwards from over 350 ft (107 m) to less than 150 ft (46 m) with an associated thinning and deterioration of the coal seams. The main workable seams are the Bench, Double and Deep Rider in the north of the coalfield, and the Seven-Feet. Two marine bands, the lower one probably the *G. subcrenatum* Marine Band, occur in the lowest beds in the north of the Coalfield. Elsewhere the Lower Coal Measures rest unconformably on Old Red Sandstone or Cambrian strata.

In South Derbyshire the Lower Coal Measures are about 1200 ft (366 m) thick, and in Leicestershire about 700 ft (213 m) thick; the sequences in these coalfields can now be correlated quite well. The lowest workable seam in each coalfield is the Kilburn, the lowest 600 ft (183 m) in South Derbyshire and 300 ft (91 m) in Leicestershire being unproductive. These lower beds contain at least three marine bands in South Derbyshire, and of these the *G. subcrenatum* Marine Band and probably the Alton Marine Band are also represented in Leicestershire. A thick sandstone, the Wingfield Flags, commonly red-stained, occurs below the Kilburn seam in both coalfields, and there is usually a thick sandstone above the Eureka Coal of South Derbyshire. The chief workable seams in South Derbyshire (with the probable Leicestershire equivalents in brackets) are the Kilburn (Kilburn), Stanhope (Low Main), Eureka (Yard), Stockings, Woodfield (Middle Lount), Little Woodfield and Main (Upper Main). These seams have an aggregate thickness of about 40 ft (12 m) in parts of South Derbyshire. Mussel bands are well developed above the Clod and Woodfield (Middle Lount) coals.

**Middle Coal Measures**

The Middle Coal Measures of North Staffordshire are about 2000 ft (610 m) thick, and are dominantly argillaceous; at the base the Seven-Feet Banbury Marine Band is well developed and contains a fauna including *Anthracoceras vanderbeckei*. This marine band has been found in all the other coalfields of Central England except the Forest of Wyre (Plate VII). The top of the Middle Coal Measures is marked by the Bay Marine Band, an horizon also present in the Cannock area of South Staffordshire. The lower part of the Middle Coal Measures contains many thick seams, but above the Moss Coal workable seams are less common. Ironstones were formerly of economic importance, particularly those associated with the Burnwood and Winghay seams. The upper part of the Middle Coal Measures contain eight marine bands which make it possible to attain a degree of precision in correlation not yet feasible in the lower beds. The Gin Mine Marine Band carries a rich fauna, and its equivalents are known from all the Midland coalfields except Leicestershire (Plate VII). Rich faunas of mussels occur in the shales above many of the seams, and above the Longton Hall and Priorsfield marine bands.

In South Staffordshire the Middle Coal Measures thin southwards from about 900 ft (274 m) to about 300 ft (91 m) accompanied by a coalescing of some of the seams similar to that in the Lower Coal Measures (Fig. 8B). The Thick Coal of the Dudley area results from the coalescing of several seams of the northern area. South of Halesowen the coals die out and the sequence consists mainly of fireclay. Working collieries are now restricted to the northern part of the coalfield, where the principal seams worked are the Park, Eight Feet, Brooch and Bottom Robins. Of the six marine bands in the upper part of the sequence in the Cannock area, all of which can be correlated with North Staffordshire, only two (the Charles and Sub-Brooch) extend into the southern part of the coalfield. Red beds of Etruria Marl facies enter above the Charles Marine Band in the southern area and eventually replace it.

A maximum thickness of about 450 ft (137 m) of Middle Coal Measures is preserved in the Coalbrookdale Coalfield, of which the top 200 ft (61 m) are largely of Etruria Marl facies. Locally, pre-Upper Coal Measures erosion has removed the whole of the Middle Coal Measures (p. 50 and Fig. 11). The main workable seams are the Flint, Double and Top. The base of the Middle Coal Measures is marked by the Pennystone Marine Band, and two higher marine bands, the Blackstone and Chance Pennystone, correlate with the Sub-Brooch and Charles marine bands of South Staffordshire. Ironstones associated with the Pennystone and Chance Pennystone marine bands were formerly of great economic importance.

The thickness of the Middle Coal Measures of the Forest of Wyre Coalfield is uncertain, since no representative of the *A. vanderbeckei* Marine Band has yet been recognized. The only known marine horizon is the Eymore Farm Marine Band, equivalent to the Charles Marine Band of South Staffordshire. The chief workable seam is the Highley Brooch. The top 300 ft (91 m) of beds are largely of Etruria Marl facies, and similar red beds form the bulk of the measures below the Four Feet Coal.

In Warwickshire, the Middle Coal Measures thin southwards from about 500 ft to 300 ft (152-91 m). As in South Staffordshire, this thinning is accompanied by a coalescing of some of the seams to form the Thick Coal of the southern part of the coalfield. The principal seams worked are the Two Yard of the central part of the coalfield, and the Thick in the south. The Seven-Feet Marine Band marks the base of the sequence, the only other marine band being the Nuneaton (Charles). The base of the Etruria Marl facies is usually within 100 ft (30 m) above the Nuneaton Marine Band, but locally red beds extend below this horizon.

Despite the short distance between the South Derbyshire and Leicestershire coalfields, it is only recently (Spink 1965) that the Middle Coal Measures sequences have been correlated in any detail. The beds are about 1200 ft (366 m) thick in South Derbyshire and 550 ft (168 m) thick in Leicestershire. As well as the Molyneux Marine Band (Bagworth of Leicestershire) at the base, at least three other marine bands are known from the South Derbyshire sequence (Plate VII). Recently, a marine band has been recorded (Horton 1963) from the highest Coal Measures in Leicestershire. The Little, Yard

and Block are the principal seams worked in South Derbyshire, and the New Main, Splent, Five Feet and Minge in Leicestershire. Several of the seams have well-developed mussel bands in the roof shales.

**Upper Coal Measures**

In Central England, Upper Coal Measures are present in all but the Cheadle and Leicestershire coalfields. They reach their maximum development, of nearly 5000 ft (1524 m) in the Potteries Coalfield (North Staffordshire), where the upward sequence comprises lower grey beds (including the Blackband Group) about 1100 ft (335 m), Etruria Marl Group about 1100 ft (335 m), Newcastle-under-Lyme Group about 400 ft (122 m) and Keele Group about 2200 ft (670 m) (Fig. 10).

*Blackband Group and lower beds.* In North Staffordshire, around Hanley, rather more than 1000 ft (305 m) of beds of 'normal' grey facies overlie the Bay Marine Band. A group of thick coals lies near the middle of these beds and the Peacock, Spencroft, Great Row and Cannel Row seams are of considerable economic importance. Below these coals, the seams are thin and of little value, though some of the associated clay-ironstones were formerly worked. In the upper 350 ft (107 m) or so between the Bassey Mine and the Blackband, several of the coals are associated with blackband ironstones and this part of the sequence has been called the *Blackband Group.* Several major collieries were initiated essentially as ironstone mines, but that industry is now defunct. Two thin, but distinctive and persistent limestones lie between the Bassey Mine and the Hoo Cannel. No marine bands are known in the Group, but mussels of the *Anthraconauta phillipsii* group together with ostracods are extremely abundant at several horizons, particularly within the blackband ironstones. Fish bands are numerous, especially below the Great Row, while bands with '*Estheria*' lie close above and below the Hoo Cannel. There are red mudstones in several of the rhythms of the Blackband Group. In the north of the coalfield, interdigitation of red and grey mudstones sets in about 100 ft (30 m) above the Blackband, and, continuing for almost 200 ft (61 m), forms a transition upwards to the Etruria Marl Group. To the south this change begins at lower horizons and between Stone and Stafford the Etruria Marl facies descends almost to the base of the Upper Coal Measures.

The Blackband Group is not present in any other of the Midland coalfields and the only other occurrence of grey measures at the base of the Upper Coal Measures is in the northern part of the South Staffordshire Coalfield where about 200 ft (61 m) of grey beds underlie the Etruria Marl.

*Etruria Marl Group.* Although its base is hard to define, the Etruria Marl forms a distinctive group throughout the Potteries Coalfield. It thins gradually southwards from about 1000 ft (305 m) to 700 to 800 ft (213–244 m). Large-scale rhythmic sedimentation is apparent throughout, commencing with a red seatearth, which may be capped by a greyish green, sometimes highly carbonaceous, bedded mudstone. The overlying bulk of the rhythm comprises red sandy mudstone with rare bands of 'espley' grit. The 'espleys' tend to be thicker and more numerous in the south. Only one coal is recorded and this appears to die out southwards. Two thin bands of *Spirorbis* limestone occur locally. Poorly preserved plants, scattered mussels (*Anthra-*

*conauta*) and ostracods occur at a few horizons, especially in the upper part of the Group.

In the other Midland coalfields red beds of similar facies to the Etruria Marl Group occur at lower horizons and are not confined to the Upper Coal Measures. This Etruria Marl facies is very variable in thickness, due partly to an unconformity between it and the overlying Halesowen Beds or their equivalents (the 'Symon Fault' of Coalbrookdale, Wyre Forest and South Staffordshire) and, in some areas, due partly to a lateral passage of the red measures into grey Coal Measures. In South Staffordshire the Etruria Marl facies is up to 700 ft (213 m) thick. In the northern part of the Coalfield its base is about 200 ft (61 m) above the base of the Upper Coal Measures, but farther south red beds enter at a lower horizon, immediately above the Charles Marine Band. 'Espleys' are well developed in the southern part of the coalfield, where they contain much angular debris of local rocks. The facies is thinner in Warwickshire (0–325 ft), South Derbyshire (about 150 ft) and Coalbrookdale (0–200 ft) and its base maintains a fairly constant horizon usually within 100 ft of the Nuneaton (Charles) Marine Band and its equivalents. In the Forest of Wyre Coalfield the basal 300 ft (91 m) of the Coal Measures are of Etruria Marl facies, and much of this sequence may be of Lower Coal Measures age. Similar beds occur at many horizons above the Highley Brooch coal (Plate VII).

*Newcastle-under-Lyme Group.* The Newcastle-under-Lyme Group of North Staffordshire overlies the Etruria Marl Group and represents a reversion to 'normal' Coal Measures sedimentation. The Group varies in thickness between about 300 and 500 ft (91–152 m), comprising rhythmic alternations of mudstone and sandstone with about six thin coals. *Anthraconauta tenuis*, '*Estheria*' and ostracods are locally abundant in the mudstone roofs of these coals. Thin limestone beds are widely developed at the base of the Group. Towards the southern part of the coalfield red beds appear in the middle of the Group and again towards its top, rendering its distinction from the Keele Group a matter of some difficulty.

The equivalents of the Newcastle-under-Lyme Group in the other Midland coalfields are probably the Halesowen Beds of South Staffordshire, Warwickshire and South Derbyshire, the Coalport Beds of Coalbrookdale and the Highley Beds of the Forest of Wyre. Lithologically they resemble the Newcastle-under-Lyme Group, and vary in thickness from about 400 to 500 ft (122–152 m) except in South Derbyshire where they are only about 150 ft (46 m) thick. The thin coal seams in these beds are of poor quality, but have been worked locally in the past, particularly in the Forest of Wyre. Persistent thin beds of *Spirorbis* limestone are useful locally for correlation purposes.

*Keele Group.* The base of the Keele Group in North Staffordshire is most readily taken at the base of a series of red sandstones. Between these sandstones, and for 200 to 300 ft (61–91 m) above them, cyclic sedimentation is apparent, and in the north, some of the intervening mudstones are grey, with thin coals. Above this sequence there are about 1000 ft (305 m) of almost unbedded red mudstones with only thin bands of sandstones and siltstones. Higher in the sequence sandstones again become important, and it is possible that a total of 3000 ft (914 m) of the Keele Group is preserved though this may include strata assigned farther south to succeeding groups. Many of the mudstones are calcareous, and beds of pellet-mudstone are common. Three or four thin porcellanous limestones are known to occur. *Spirorbis* ranges throughout the Group, and '*Estheria*' is present near the base. The mussel *Anthraconaia* occurs sporadically and the small gastropod *Anthracopupa* is known from the middle of the sequence. The Keele Group is of similar lithology in the other Midland coalfields, and varies in thickness from about 100 ft (30 m) in South Derbyshire to 900 ft (274 m) in Warwickshire and South Staffordshire.

### Igneous Rocks in the Coal Measures

Igneous rocks (mainly basalts and dolerites) occur in association with Coal Measures at several localities in Central England. They can be considered in two groups; firstly, those lying within a narrow belt of country which extends from the Clee Hills to the Leicestershire Coalfield, and secondly, subsurface occurrences at the southern end of the Yorkshire–East Midlands Coalfield (The Pennines and Adjacent Areas, *Brit. Reg. Geol.*).

The first group includes the igneous rocks of the Clee Hills, Kinlet, Shatterford, South Staffordshire and Whitwick. Many of these bodies were at one time considered to be intrusions of Tertiary age, but are now believed to date from the late Carboniferous or early Permian. Pocock (1931) described the field relations of most of them and concluded that several of them were lava flows. However, Marshall (1942), on re-examination of the evidence regarded them as intrusions (with the possible exception of the Kinlet Basalt), and this is supported by later palaeomagnetic evidence (Everitt 1960).

The basalt of Titterstone Clee Hill (Plate VA), which may be as much as 300 ft (91 m) thick in places, lies chiefly within the Coal Measures, although in the north-western part of its outcrop it appears to rest on Old Red Sandstone. Near Cornbrook it is about 180 ft (55 m) above the Great Coal. Small areas of basalt, resting on Coal Measures, form the summits of Brown Clee Hill.

The Kinlet Basalt, in the Forest of Wyre Coalfield, occurs at about the horizon of the Highley Brooch Coal. It has variously been regarded as a sill or lava-flow, and Marshall (1942) considered that the available evidence was inconclusive. Later palaeomagnetic data (Everitt 1960) suggests that it is intrusive. At Shatterford, also in the Forest of Wyre, a basalt occurs in steeply dipping Coal Measures. Pocock (1931) described it as a lava flow on account of the absence of alteration of the overlying sediment, its vesicular top and the occurrence of pipe-amygdales at the base. Marshall (1942), however, considered it to be an intrusion, since the basalt fingers into the overlying shales, and this is again supported by palaeomagnetic evidence. Igneous rocks have been proved underground in the Highley area, and also in the Claverley Borehole between the Forest of Wyre and South Staffordshire coalfields.

Several intrusive masses of basalt occur in the South Staffordshire Coalfield. They crop out in the Rowley Hills, at Pouk Hill near Walsall, at Wednesfield and at Barrow Hill near Pensnett, and have also been proved in many shafts, borings and opencast sites, particularly in the Landywood area, about three miles south-south-east of Cannock, where a sill reaches a maximum thickness of 180 ft. The Rowley basalt has the general form of a laccolithic intrusion into the Etruria Marl (Whitehead and Eastwood 1927). It has a maximum thickness of over 100 ft (30 m) and at its margins passes into sills a foot or so in thickness. The rock, known locally as 'Rowley Rag', shows well-defined jointing, usually of columnar type. The Wednesfield and Pouk Hill basalts are intruded into Lower Coal Measures. Where these various basalts cut through coal seams, the coal is sometimes destroyed for some distance on either side, or may be converted to the condition of coke or anthracite. Nevertheless, contact metamorphism is very limited in extent.

Several sills of dolerite with an aggregate thickness of more than 80 ft (24 m) occur in the highest beds of the Middle Coal Measures, immediately below the Triassic, over an area of nearly six square miles (16 km$^2$) around Whitwick and Ellistown collieries in the Leicestershire Coalfield. A recent age determination (Meneisy and Miller 1963) suggests that the dolerite may be of mid-Permian age.

The second group of igneous rocks, which have been proved in boreholes for oil and coal (Falcon and Kent 1960) at the southern end of the Yorkshire–East Midlands Coalfield, includes both extrusive and intrusive types. Extrusive rocks occur in two areas, at different levels within the Lower Coal Measures. To the west and south-west of Newark-on-Trent, flows of albitized olivine-basalt, with a total thickness of up to 60 ft (18 m), occur near the top of the Lower Coal Measures. Farther south, around Screveton, Plungar and Sproxton (Fig. 6), igneous rocks, probably mainly extrusive, locally reach a thickness of nearly 500 ft (152 m) within the lower part of the Lower Coal Measures. Intrusive rocks, chiefly sills of micro-teschenite and micro-gabbro up to 100 ft (30 m) thick, occur in the basal Lower Coal Measures over a wide area from Newark-on-Trent southwards to near Melton Mowbray. They are probably of post-Triassic origin since the Coal Measures sequence and the unconformity between the Permo-Triassic and the Carboniferous are both displaced upwards over the intrusions.

# 7. Permo-Carboniferous

The Enville Beds, a series of continental red beds, are generally grouped with the Upper Coal Measures. They appear to succeed the Coal Measures without any major stratigraphical break, but may be in part of Permian age. The problem of correlation is accentuated by the general absence of fossils and by the effects due to earth movements which were contemporaneous with their deposition. Evidence of the possible Permian age of at least part of the Enville Beds is the presence of the skull of the labyrinthodont *Dasyceps bucklandi* and the maxilla of the reptile *Oxyodon* in the Kenilworth Sandstones.

Although the Enville Beds usually succeed the Keele Group without marked unconformity, they were deposited over a more restricted area. They are exposed in the Wyre Forest, Coalbrookdale, South Staffordshire and Warwickshire coalfields and have probably been encountered in boreholes near Lichfield and Stratford-upon-Avon.

The scattered character of the outcrops has resulted in the introduction of many local formational names. Generally the strata consist of red 'marls' and red calcareous sandstones, with locally thick beds of calcareous conglomerate in the lower part and of breccia in the upper part of the sequence. In South Staffordshire they are divided into a lower or 'Calcareous Conglomerate Group' and an upper or 'Breccia Group'. Wills (1956, p. 84) has expanded this into a tripartite classification which is applicable throughout the Midlands: the lowest division includes the Corley, Bowhills and Calcareous Conglomerate Groups, the middle division contains the Tile Hill Group and the upper division includes the Gibbet Hill Group and the Ashow and Kenilworth Breccias (and sandstones) of Warwickshire, and the Haffield, Abberley, Enville, Clent and Nechells Breccias.

The Enville Beds accumulated under arid conditions as evidenced by the presence of pellet rocks, footprint beds, rainspot impressions and suncracked surfaces at several horizons. The deposits are the products of the erosion of the Mercian Highlands and were laid down in an irregular series of interconnecting basins. They attain a maximum thickness of 3500 ft (1067 m) in Warwickshire but thin westward to 1000 ft (305 m) in South Staffordshire and to 700 ft (213 m) in the Wyre Forest. While they were being laid down, the Mercian Highlands and Welsh Uplands were undergoing periodic elevation and the conglomerates and breccias which occur at intervals throughout the succession may mark periods of uplift when erosion increased and the transporting power of rivers was greatly enhanced. It is in the marginal areas close to the rising mountains that uplift and erosion resulted in the development of unconformities. Thus in the Wyre Forest and in the southern part of the South Staffordshire Coalfield an unconformity has been traced at the base of the Breccia Group. This unconformity appears to be relatively widespread but it does not extend into the Warwickshire area where no important break is known, either at the base of or within the Enville Beds.

**The Calcareous Conglomerate Group, Bowhills Group and Corley Group**

The Enville conglomerates are largely composed of pebbles of limestone and chert of Carboniferous age mixed with pebbles of older rocks, the relative proportions varying considerably from place to place. The material appears to have been of local derivation. Conglomerates near Abberley contain pebbles of the dolomitic type of Wenlock Limestone which is peculiar to that district, whereas contemporary conglomerates in South Staffordshire contain pebbles of the Dudley type of Wenlock Limestone, and also of the local Cambrian quartzite. From the high proportion of Carboniferous Limestone pebbles in conglomerates in South Staffordshire and the Enville district, it is concluded that Carboniferous Limestone cropped out over large tracts of the Mercian Highlands between the Wyre Forest and South Staffordshire areas, and probably again to the east of Birmingham.

Shotton (1929) has traced five conglomeratic horizons in the Enville Beds of the Warwickshire Coalfield. The lowest three of these, the Arley (Exhall), Corley and Allesley Conglomerates are thought to lie within the Calcareous Conglomerate Group division. The Corley Conglomerate is the main aquifer of the Coventry district; it is largely composed of pebbles of Llandovery Sandstone derived from a former outcrop of that formation to the east. In the Coventry district a progressive increase in the proportion of pebbles of pre-Carboniferous rocks occurs as the beds of conglomerate are traced in upward sequence: this indicates that as erosion of the mountain ranges proceeded, successively older rocks were exposed to denudation. The conglomerates occur in lens-shaped masses associated with sandstones and 'marls' (silty mudstones) and represent gravels laid down on piedmont fans which spread out and coalesced at the foot of the Mercian Highlands. Away from the source area the pebbles are less frequent and the environment may have been deltaic.

**Tile Hill Group**

These beds are present in Warwickshire where they consist of about 900 ft (274 m) of red 'marls' (silty mudstones) with subordinate sandstones and a single conglomerate, the Beechwood Conglomerate. The limited outcrop and lack of coarse sediments suggests that they were deposited in a playa or temporary lake within an intermontane basin in an area of low relief.

**The Breccia Group**

In Warwickshire, the Tile Hill Group is conformably overlain by a thick arenaceous group with minor silty mudstones ('marls'), a thin conglomerate (the Gibbet Hill Conglomerate) and two beds of breccia. These beds have been described collectively as the Kenilworth Sandstones and have been subdivided into the Gibbet Hill Group, the Kenilworth Breccia Group and at the top the Ashow Group. The conglomerate in the Gibbet Hill Group has a maximum thickness of 50 ft (15 m) and passes westwards into beds of breccia. The conglomerate and the overlying breccias thin and disappear when traced eastwards, which suggests that they were derived from uplands which lay to the west of Kenilworth. The conglomerate contains a high proportion of Silurian and Carboniferous Limestone pebbles but the overlying breccias show a progressive increase in the percentage of Pre-Cambrian rocks. At the base of the Ashow Group there is a predominantly argillaceous group which has been dug for brick-making and from which footprints and plants have been recorded.

Elsewhere the Breccia Group consists of breccias interstratified with 'marls' and sandstones. The breccias contain angular fragments of rock, with a glazed coating of hematite, in a 'marly' matrix. Most of the fragments are from 2 to 6 inches (5–15 cm) in diameter, but blocks up to 2 ft (0·6 m) across have been recorded. Some of the associated finer sediments and the matrix of the breccias consist of small rock fragments and rock flour. The lack of rounding, together with the large size of the pebbles suggest that the breccias were rapidly deposited by torrential streams. Wills (1956, p. 100) suggests that the angular material originated as screes formed against a series of uplifted fault blocks. The environment may be compared with that of extensive breccia-fans forming at the present day in the desert plains of Peru.

Whilst the underlying conglomerates consist mainly of Carboniferous and Silurian rocks, the breccias are composed of fragments of rocks older than the Silurian, particularly of Pre-Cambrian rocks. Because of the high percentage of pebbles of Uriconian volcanic rocks in the breccias of the West Midlands, they are often referred to as the 'trappoid' breccias. The thickest development, about 400 ft (122 m), is the Clent Breccias near Birmingham. The abundance of Uriconian rock-fragments in these indicate that outcrops of Uriconian rocks formerly existed between the Wyre Forest and South Staffordshire.

The Enville Breccias are highly resistant to erosion, and outcrops form well-defined ridges, such as the escarpment of the Clent Hills to the south-west of Birmingham. They are well developed in the Wyre Forest Coalfield, cropping out between Enville and Claverley, at Bowhills, and along the line of the Eastern Boundary Fault, capping Castle Hill, and again at Warshill in the Abberley Hills. They crop out along the western margin of the South Staffordshire Coalfield, for example, in the Lickey, Clent and Walton hills, and farther east in the same coalfield at Frankley, Northfield and Warley.

Fossils are extremely rare in the Enville Beds; plant remains have been found at several levels in the Corley Beds of Warwickshire, and include impressions of leaves of *Walchia* and *Asterophyllites*, and silicified wood of *Dadoxylon* and *Cordaites*. Footprints of amphibia have been found at several localities, for example at Hamstead, and in the Whitemoor Brick works near Kenilworth.

# 8. Permo-Triassic

In the Midlands it is impossible to define precisely the position of the boundary between the Permian and Triassic systems, and the general term 'New Red Sandstone' is still employed as a means of describing the Permo-Triassic rocks (excluding the Rhaetic) of Central England. In spite of the extensive nature of the outcrop of the New Red Sandstone, beds of undoubted Permian age are limited in their distribution being mainly restricted to the northern extremities of the region. The Permo-Triassic rocks form the relatively low-lying but undulating country of the Midland and Cheshire plains.

The paucity of fossils in the Permo-Triassic rocks of Central England has led to a classification based on lithology:

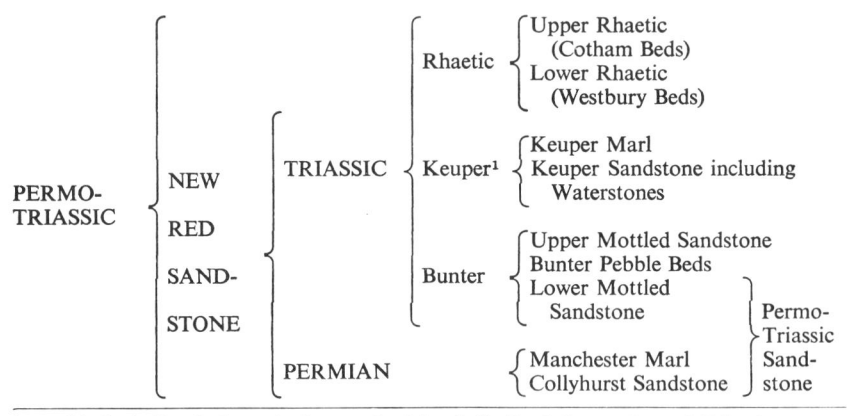

| PERMO-TRIASSIC | NEW RED SAND-STONE | TRIASSIC | Rhaetic | Upper Rhaetic (Cotham Beds) / Lower Rhaetic (Westbury Beds) | |
|---|---|---|---|---|---|
| | | | Keuper[1] | Keuper Marl / Keuper Sandstone including Waterstones | |
| | | | Bunter | Upper Mottled Sandstone / Bunter Pebble Beds / Lower Mottled Sandstone | Permo-Triassic Sandstone |
| | | PERMIAN | | Manchester Marl / Collyhurst Sandstone | |

[1]Possibly including beds equivalent to the Muschelkalk of the continental Triassic.

These divisions are lithological formations or 'rock-stratigraphic' units and the age of each division may not be the same throughout the area.

The Hercynian earth-movements reached their climax in the Midlands soon after the deposition of the Coal Measures. The older rocks then formed a mountainous tract of folded and faulted rocks considerably exceeding in extent the Mercian Highlands of the Carboniferous Period (pp. 41, 57). These mountains were subjected to intense denudation, the magnitude of which can be seen throughout the Midlands, where the basal beds of the New Red Sandstone rest with strong unconformity upon the Carboniferous and older rocks. By the middle of Permian times, the relief had been greatly subdued: in the north-eastern part of the region the Zechstein Sea transgressed westwards across a low-lying plain which probably terminated against a low barrier ridge which existed in the present Pennine region; in the south the highlands had been reduced to a series of low mountain ranges separating a series of intermontane basins, in which products of erosion were still accumulating.

Many of the present day Triassic basins are defined by faults, some of which may initially have been of pre-Triassic age. Their synclinal form may have originated at the time of deposition but is largely due to warping and faulting during post-Keuper earth movements. In the areas where continuous relative subsidence took place great thicknesses of continental beds accumulated. The maximum development of New Red Sandstone is in Cheshire and North Shropshire where an estimated thickness of 8600 ft (2621 m) of strata are preserved in the axial region of the Cheshire basin. Other major developments occur in the part of the Severn Basin which lies within the Central England Region, and in North Staffordshire. Elsewhere the thickness is extremely variable, at least 1500 ft (760 m) being present north of Bridgnorth, 1500 ft (460 m) at Birmingham and 500 ft (150 m) below the Jurassic. The Triassic deposits thin out against the London Platform (the remnant of the Mercian Highlands) and today they are absent beneath the Jurassic south-east of a line running from Oxford to the vicinity of Norwich. Borings at Orton, Oxendon and Gayton have each proved less than 100 ft (30 m) of sediments of Keuper type.

The Permo-Triassic rocks are predominantly red in colour, due to the presence of hematite with small amounts of goethite. In sandstones the iron oxides usually form a pellicle on each grain, the intensity of the body colour being related to the thickness of this layer. The presence of the pellicle on well-rounded or 'millet-seed' sand grains in deposits considered to be of aeolian origin suggests that it was precipitated from ferruginous solutions percolating through the rock. The amount of iron rarely exceeds 5 per cent, the maximum being found in argillaceous deposits where the iron minerals were included at the time of deposition. Oxidizing conditions usually persisted in the depositional environment, but, occasionally, green beds may have resulted from the development of reducing conditions or from the diagenetic reduction of the ferric iron colouring. The colouring matter was derived from the source area of the sediments and was probably produced by lateritization under warm climatic conditions. The preservation of the red colour in the rocks reflects the lack of humic material within the basin.

In Central England the New Red Sandstone consists largely of sandstones, conglomerates and 'marls' (quartzose mudstones and siltstones). The abundance of wind-worn or 'millet-seed' quartz grains, especially in the lower divisions, suggests relatively arid desert conditions. The Bunter Pebble Beds contain evidence of rhythmic deposition in water, each rhythm commencing with a conglomeratic sandstone which passes upwards into 'marl'. Major slump structures occur in the Upper Mottled Sandstone showing that parts of this aeolian formation were deposited in water. By the commencement of Keuper times much of the old land surface had been eroded to base level so that the fluviatile and deltaic sandstones of the basal Keuper extend over much wider areas than the earlier Permo-Triassic formations. By late Keuper times most of the sediment being laid down was of clay and silt grade. During this latter phase wide shallow lakes came into existence and some of these may have been subjected to incursions of sea-water. The absence of free drainage, and the high evaporation, led to

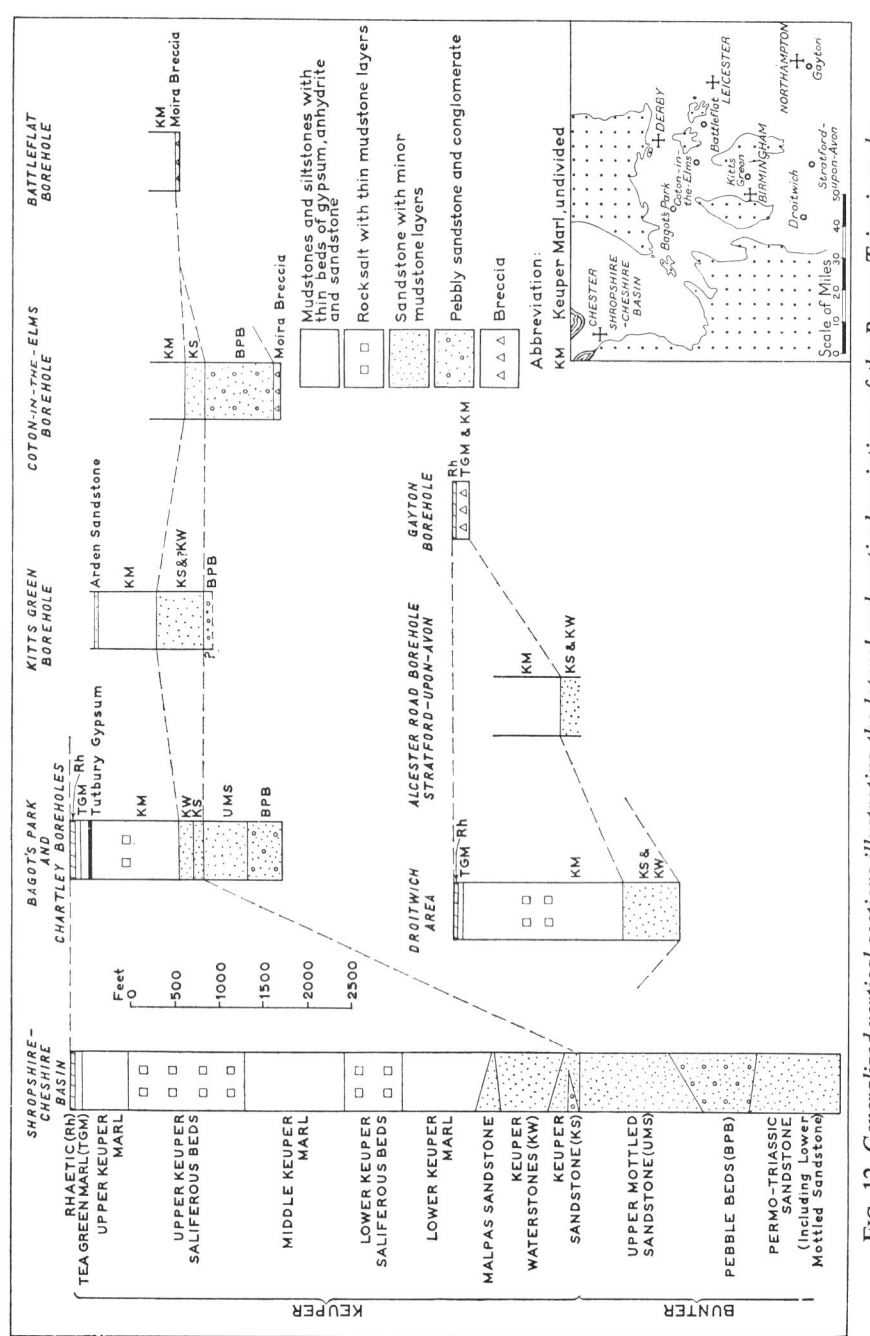

FIG. 12. Generalized vertical sections illustrating the lateral and vertical variations of the Permo-Triassic rocks
In the inset map the outcrop of the Carboniferous and older rocks is stippled

increased salinity and to the creation of an environment unsuitable to life in which the deposition of rock salt, gypsum and anhydrite took place.

The Permo-Triassic rocks are poor in organic remains. The bones and footprints of terrestrial animals are the most prominent fossils, and carapaces of the bivalved crustacean *Euestheria* are occasionally found. At the present day a similar conchostracan lives in fresh or brackish water and is capable of surviving wide ranges of temperature and extreme drought.

## Permian

Marine Permian rocks do not crop out in Central England, though boreholes in the Vale of Belvoir have proved Carboniferous rocks overlain by a basal breccia, Lower Permian Marl and Lower Magnesian Limestone. These beds die out southwards, either being replaced by pebbly sandstones indistinguishable from the overlying Bunter rocks or cut out beneath the Bunter unconformity. In the South Derbyshire Coalfield, the basal breccia of the New Red Sandstone (the Moira Breccia), was once thought to be of Permian age: the breccia is diachronous and is developed at the base of the Keuper Marl in Leicestershire where this formation has overstepped the older formations.

## Triassic

The variation in the thickness of the Triassic rocks is largely the result of subsidence in the major depositional basins, but, particularly in the case of the lowest beds, the infilling of hollows in the pre-existing land surface was an important factor. Some of the variations may be due to banking of sediments against fault scarps which were active during the period of deposition (Tonks *et al.* 1931; Wills 1935), but in several cases it has been shown that the isopachytes of these formations are unrelated to the orientation of the fault planes (Poole and Whiteman 1955).

### Bunter

The triple sub-division of the Bunter, the lower part of the Trias, is best seen near Bridgnorth in Shropshire, where the Series attains a thickness of up to 1500 ft (460 m). The three sub-divisions persist throughout that county and through much of Cheshire, Staffordshire and Worcestershire.

*Lower Mottled Sandstone.* This, the lowest division of the Bunter, is regarded by some as being of Permian age. The bulk of the formation is composed of fine-grained sandstone with local marly bands, bright red in colour with occasional streaks and patches of yellow sometimes having a greenish tint. These sandstones are highly sorted and contain an abundance of well-rounded wind-polished quartz grains with small proportions of feldspar and mica. Some of the sandstones show large-scale cross-stratification and many of them are thought to have been deposited as dune sands. The orientation of the dips of the cross-stratification suggests that the prevailing wind blew from the east. This facies, known as the Dune Sandstone Group, is well developed near Bridgnorth where the basal beds include a conglomerate.

In Shropshire the Lower Mottled Sandstone varies in thickness from 80 ft to about 600 ft (24–183 m). Near Kidderminster contemporaneous fault movements are thought to have controlled deposition and to have resulted in rapid local variations in thickness. In the Wolverhampton area the formation is about 1000 ft (305 m) thick; it thins eastwards and has not been recorded in the Burton-upon-Trent area. A thin group of sandstones of Bunter type are developed in South Nottinghamshire, but the lowest beds are of Permian age. In the Manchester area the Permian Collyhurst Sandstone and marine Manchester Marl are developed. But when traced south to the Central England Region the argillaceous beds pass laterally into sandstones which are indistinguishable from the Collyhurst Sandstone below and the Bunter Sandstone above. In Cheshire the group was classified as 'Lower Mottled Sandstone' and more recently as the Permo-Triassic Sandstone. The latter varies in thickness from 200 to 1000 ft (67–335 m), the lower and greater part being equivalent to the Collyhurst Sandstone. In this thick sandstone it is impossible to draw a line between Permian and Trias and the formation is therefore discussed with the Bunter with which it is traditionally grouped. Although 'millet-seed' type sand grains abound in these rocks, the sandstones may have accumulated in water.

In several areas in the Midlands breccias underlying the Pebble Beds rest directly on Carboniferous rocks, and they may be in part the local equivalent of the Lower Mottled Sandstone. They include the pebbly Barr Beacon Beds, the quartzite breccias of South Staffordshire, the Hopwas Breccia and in South Derbyshire parts of the Moira Breccia. They are similar in character to the Enville breccias (p. 58) with which they have sometimes been grouped, but Eastwood (*in* Eastwood *et al*. 1925) has shown that the Enville Beds have been affected by earth-movements which are earlier than those affecting these breccias. Therefore they are of post-Enville age and are probably scree deposits of early Triassic age.

*Bunter Pebble Beds*. These consist of coarse-grained, brownish red current-bedded sandstones with conglomerate lenses and layers of pebbles. The pebbles are well rounded and vary in size from $\frac{1}{4}$ in (7 mm) to about 9 inches (23 cm) in diameter, a few being larger. In some places pebbles are sparingly scattered through the sandstones, elsewhere they may be in bands or large lenticular masses. On Cannock Chase the pebbles are so abundant that they form poorly cemented sandy conglomerates readily weathering to masses of loose pebbles. Near Bridgnorth the beds have a calcite cement whilst in South Derbyshire gypsum-rich bands are present.

The Pebble Beds may be as much as 1000 ft (305 m) thick in the Cheshire Basin, where they consist of sandstones with conglomerate beds and, more rarely, bands of sandy mudstone. The association of slump structures with cross-stratification and current-bedded lenticular conglomerates is evidence of deposition in water.

The Pebble Beds are more widespread than the Lower Mottled Sandstone and in parts of Staffordshire rest unconformably upon older rocks. They extend into South Derbyshire where they rest upon the basal Moira Breccia. Here they fill in a steep-sided depression which is said to be the valley of the Triassic 'Polesworth' River (Wills 1950). Around Bridgnorth the Pebble Beds are between 370 and 400 ft (113–121 m) thick and rest upon an eroded

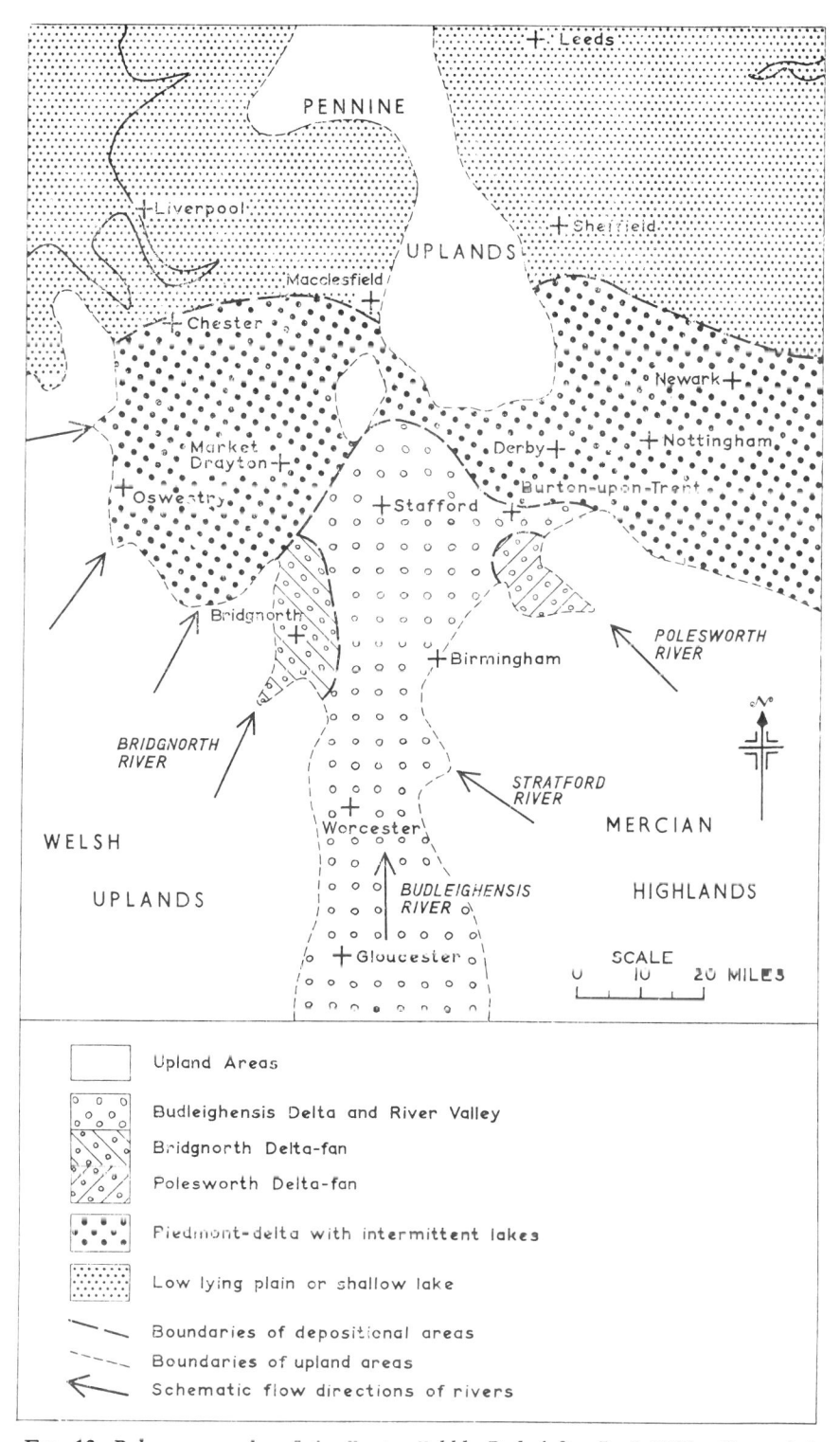

FIG. 13. *Palaeogeography of the Bunter Pebble Beds* (after L. J. Wills, Concealed Coalfields)

surface of Lower Mottled Sandstone. Locally the lowest bed is a calcareous breccia or coarse grit, elsewhere a calcareous conglomerate is present, whilst lenticular breccias may occur in the beds above.

The formation appears to have originated as desert-fan and valley-fill deposits formed by intermittent torrential streams flowing northwards from the Mercian Highlands. Traced northwards from Bridgnorth the decrease in grain size of the Bunter sediments was probably associated with a change in the depositional environment to one of low-lying piedmont delta and finally to an area of temporary lakes. Probably the building of each fan was relatively rapid.

The pebbles include a variety of rock types, but the relative proportions are characteristic in certain areas. Thus several deltas have been recognized, e.g. grey and liver-coloured quartzites and vein-quartz pebble suites characterize the 'Budleighense' River (Wills 1950) which flowed northward in the Birmingham area. The provenance of the quartzite is uncertain, but suggested sources range from Brittany to Scotland. However, southerly derivation is favoured since the pebbles die out northward in Lancashire. Other far-travelled pebbles include tourmalinized rocks, possibly from southwest England, and fossiliferous quartzites, some Ordovician, some Devonian, which are characteristic of Armorica (Devon and Northern France). Undoubted local rocks such as Lickey Quartzite, Llandovery Sandstone and Carboniferous Limestone are present and are most common in the basal beds. There are also pebbles of metamorphic and igneous rocks.

*Upper Mottled Sandstone.* These beds are best developed in the West Midlands and Cheshire. They consist of soft bright foxy-red medium to fine-grained sandstone with thin beds of red mudstone. Each sand grain has a coating of iron oxide. Pebbles are almost completely absent, the formation succeeding the Pebble Beds with apparent conformity. At Peckforton, south of Tarporley, the Sandstone may be 1400 ft (427 m) thick; around Bridgnorth it is 500 ft (152 m) thick, but it thins eastward and has not been recognized beyond Birmingham. Its original extent was less than that of the underlying Pebble Beds. The sandstones are generally very feebly cross-stratified. In Cheshire they contain slump structures and it is thought that much of the formation was laid down in water, probably within a shallow lake, though parts are clearly sub-aerial.

Lenses of mudstone occur sporadically throughout the Bunter; they vary in thickness from a few inches to a foot or two (5–60 cm). At several places they have yielded the carapaces of *Euestheria*, the only animal remains found in the Bunter of Central England, apart from a single specimen of a fish from Aggborough, near Kidderminster.

**Keuper**

The basal Keuper is usually sharply differentiated from the underlying Bunter. During Keuper times the depositional basins of the Bunter were considerably extended. In the western parts of Central England the Keuper is conformable upon the Upper Mottled Sandstone when present, but elsewhere it oversteps the lower divisions of the Triassic and rests with strong unconformity upon the older rocks. This is particularly well seen in

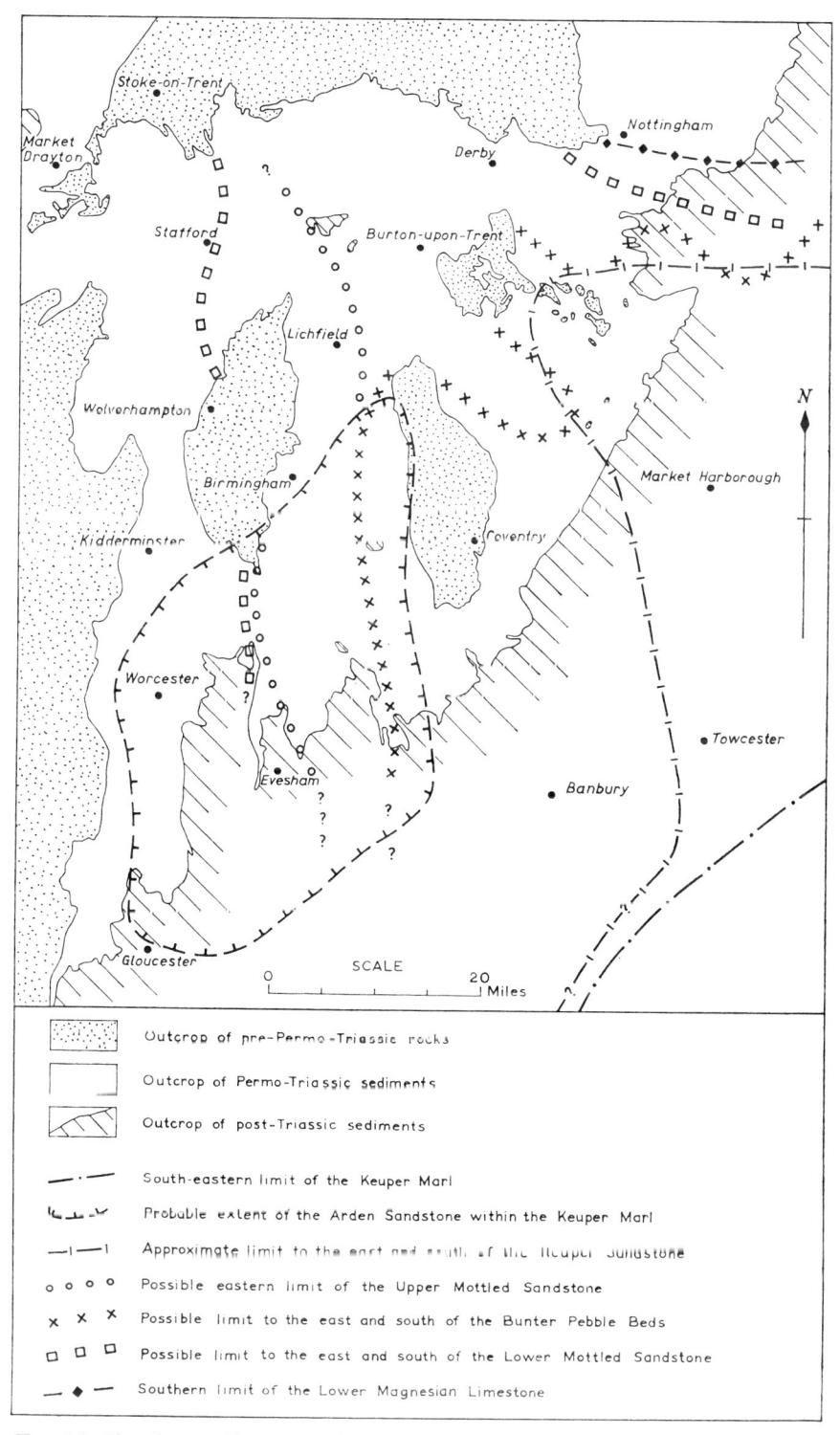

FIG. 14. *Sketch-map illustrating the existing distribution of the Permo-Triassic rocks and the progressive expansion of the depositional basin*

Charnwood Forest where the hard and resistant Charnian rocks gave rise to an upland area that was not finally buried until late in Keuper times.

Generally in the continental environment in which many red beds are laid down the maximum accumulation occurs in the peripheral areas of the depositional basins. In this region however the maximum thickness of Keuper strata occurs in the centres of the existing basins most of which were original centres of deposition. In the south-east, the Keuper thins gradually against the London Platform and is probably absent in North Buckinghamshire having been overlapped by the Jurassic.

The Keuper consists of the Keuper Sandstone overlain by the Keuper Marl; in places these subdivisions are separated by a passage group.

*Keuper Sandstone including Waterstones.* In Shropshire and parts of Staffordshire this formation varies from 100–500 ft (30–152 m) and can be divided into Basement Beds, Building Stone Group and Waterstones. Traced eastwards these divisions become increasingly difficult to recognize. In some areas including Cheshire the Basement Beds and Building Stone Group (or their equivalents) are known in combination as the 'Keuper Sandstone', the Waterstones being described as the Keuper Waterstones. The Basement Beds include conglomerates and local breccias, but the bulk of the formation consists of fine- to coarse-grained sandstone with minor bands of red and green mudstone.

In mid-Cheshire the Upper Mottled Sandstone is overlain by about 40 ft (12 m) of passage beds, consisting of interbedded sandstone of Bunter and Keuper types, succeeded by a conglomerate bed. In the Peckforton Hills, to the south of Tarporley, the Keuper Sandstone Conglomerate overlaps the passage beds and rests unconformably upon the Upper Mottled Sandstone. The bulk of the Keuper Sandstone comprises gently cross-stratified, medium to fine-grained red, brown or buff sandstones. They differ from the Upper Mottled Sandstone in being paler in colour and in the angular character of the individual sand grains. Sandstones of Upper Mottled Sandstone facies appear at at least two horizons in the Keuper Sandstone. The major development occurs near Chester, where the Frodsham Beds (Strahan 1882) form the upper part of the sequence beneath the Waterstones.

In Shropshire and Cheshire a change of lithology at the base of the Waterstones indicates an important alteration in the conditions of sedimentation. Here Keuper Sandstone is succeeded by the Waterstones, the latter being platy thin-bedded brown sandstones and siltstones, often argillaceous and separated by red mudstone layers. The origin of the name Waterstones is controversial and it has been suggested that the play of light upon the mica-laden bedding planes resembles 'watered-silk'. It is certain that the name is not meant to imply that the rock is a good aquifer. Nevertheless, when these beds crop out in steep-sided gullies it is often found that water, held up by innumerable shale partings, oozes out at numerous points on the rock face and encourages the growth of ferns and other moisture-loving plants. Salt crystal pseudomorphs, ripple marks and desiccation breccias associated with footprints of *Cheirotherium* indicate deposition in shallow water which frequently dried up. Lithologically the Waterstones in Cheshire and Shropshire are much more closely allied to the Keuper Marl than to the Keuper Sandstone, though elsewhere in the region

Plate VIII

A. The trace fossil Diplocraterian, Great Oolite Limestone, Cosgrove, Northamptonshire (A.10163)

B. Keuper Conglomerate on Keuper Passage Beds, Raw Head, Bulkeley, Cheshire (L.9)

*(For full explanation see p. viii)*

they have been regarded as passage beds from Keuper Sandstone to Keuper Marl. Recent unpublished work on the Waterstones has suggested that they may be equivalent with the Muschelkalk of the continental Trias. If this is established then it will be necessary to reclassify the Keuper Sandstone. In North Cheshire the Waterstone facies may be 900 ft (274 m) thick. Around Malpas it is succeeded by the Malpas Sandstone, at least 600 ft (183 m) of soft cross-stratified sandstone, with 'millet-seed' sand grains and infrequent argillaceous bands similar in lithology to the Upper Mottled Sandstone and the Frodsham Beds (Poole and Whiteman 1966).

In Worcestershire beds which are probably equivalent to the Keuper Waterstones are grouped with the underlying sandstones and were known as the Lower Keuper Sandstone. The lower beds are torrent-bedded sandstones with conglomerates and occasional beds of hard calcareous marl-breccia. Around Stourbridge and Bridgnorth up to 40 ft (12 m) of beds, mainly sandstones of Keuper type, separate the Upper Mottled Sandstone and the coarse conglomerates and marl-breccias. The higher beds are coarse cross-stratified sandstones with thin penecontemporaneous breccias and 'marls' and mudstones. Wills (1910 and 1947) has recorded a variety of plants and animals from these beds at Bromsgrove. Plant remains, which are usually preserved in the argillaceous bands, include the horsetails *Schizoneura* and *Equisetites*, and *Voltzia*, an early conifer. The leaves of these plants appear to have been tough and leathery like those of living species adapted to an arid environment. The fishes include forms which might have lived in rivers or pools, for example, the lungfish *Ceratodus*, present-day relatives of which are able to survive periods of drought. Derived remains of terrestrial animals include the scorpions, *Mesophonus* and *Spongiophonus*, associated with labyrinthodont amphibians and reptiles including carnivorous dinosaurs. The minute crustacean *Euestheria* is again quite common in the beds of mudstone.

Sedimentary structures recorded from the Keuper Sandstone include oscillation ripple marks, rain pittings, desiccation cracks and casts of plants, which indicate accumulation in shallow water that periodically dried up. They may be associated with flute and groove casts which Cummins (1958) suggested were formed by flood-water heavily laden with sediment which was deposited on a gently sloping mud or dust covered surface.

*Keuper Marl.* The Keuper Marl is a thick and widespread series which overlaps all the earlier beds and covers considerable areas of Worcestershire, Warwickshire, Leicestershire and Cheshire. In Cheshire, possibly the original centre of a Triassic basin, it attains a thickness of over 4500 ft (1370 m), while in Leicestershire, where it rests directly on Pre-Cambrian rocks, it has a maximum thickness of about 650 ft (198 m).

It succeeds the Waterstones conformably and many of the features noted in that formation continue throughout the Keuper Marl, the main change being a decrease in arenaceous material and an increase in mudstone and evaporite minerals, including dolomite, gypsum and, in the saliferous beds, halite. The Keuper Marl thus comprises mainly reddish brown mudstones and silty mudstones, with subordinate bands of sandstone and siltstone. These mudstones are traditionally known as 'marls', but the term may be misleading, since it is generally used to denote a highly calcareous clay, whereas the 'marls' of the Keuper, apart from the Tea Green Marl, are only slightly calcareous. They consist largely of clay minerals mixed with a high proportion of quartz dust of aeolian type, together with minute crystals of dolomite and aggregates of gypsum. Grain size analysis shows some 'marls' to have a bimodal size distribution, suggesting a combination of transporting agents, possibly wind and water. Others are very well sorted, very fine-grained and devoid of large grains of detrital sediments. These may be the result of deposition of wind-blown material in regions where intermittent lakes were formed. Taylor (*in* Taylor *et al.* 1963, p. 74) suggested that 'the banded material represents dust blown into standing water', whilst 'the blocky unstratified material may be the result of dust deposition on dry or nearly dry ground'.

Although the Keuper Marl is for the most part red or chocolate coloured, green blotches and bands give the beds a variegated appearance. 'Fish eyes' occur throughout the 'marls', these being spherical zones of decoloration which have a minute black central core, probably a radioactive or diagenetic mineral. Thin beds of grey to buff or pink sandstone, commonly associated with silty shales, occur at intervals in the Keuper Marl, particularly in the lower part. These beds, known as 'skerries', usually consist of quartz with some feldspar, worn dolomite crystals and other detrital minerals. They may be associated with 'marlstones' which consist of dolomite rhombs and subordinate calcite, associated with a variable proportion of detrital grains. The 'skerries' are often ripple marked and show sun-cracked surfaces and salt crystal pseudomorphs; fine lamination and cross-stratification are common features and may be associated with slumped bedding. In addition, groove casts, load casts, air-heave structures, raindrop imprints and worm trails have been recorded. Klein (1962) has suggested that all these structures are of lacustrine origin.

Taylor (1963, p. 72) has recognized an irregular rhythmic sedimentation within the Keuper Marl, the ideal cycle consisting of banded strata at the base, passing up into 'blocky' stratified mudstone. The lower group, consisting of dolomitic siltstones and fine-grained sandstones alternating with mudstones, contains abundant sedimentary structures. The upper part of the cycle is thought to be structureless due to reworking of the mudstone before final consolidation took place.

Skerries usually occur in groups in which sandstones alternate with 'marls' and shales; to these groups the term 'skerry belts' has been applied. In the relatively flat topography to which the Keuper Marl gives rise, almost every hilly feature is the expression of these relatively resistant sandstones, except in Cheshire and Shropshire, where thick glacial deposits effectively mask the solid rocks.

A development of thin, grey sandstones and bluish grey shales occurs in the upper part of the Keuper Marl of Worcestershire, Warwickshire and Gloucestershire. This group, the Arden Sandstone, is about 40 ft (12 m) thick, and includes coarse cross-stratified sandstones and fine-grained dolomitic sandstones. Locally it has yielded numerous fossils, notably plants, fishes, amphibia and *Euestheria*. Many of the fossils are present in a basal pellet-rock well developed near Alcester. Lamellibranchs, possibly marine, occur at Shrewley in Warwickshire. The Arden Sandstone may represent a return to the conditions of the Waterstones with periods of high rainfall moderating the arid climate which existed throughout Keuper times. The formation was deposited in shallow water possibly in a deltaic environment.

The highest beds of the Keuper Marl consist of a group of grey, green and white 'marls' which from their colour are known as the Tea Green Marl. At the base there may be a few beds of red 'marl' interbedded with the paler strata. It is uncertain whether the colour is original or secondary.

In the Charnwood Forest region sub-aerial erosion of the Keuper Marl is revealing a buried Triassic landscape; the old terrain was formed of hard Pre-Cambrian rocks and is of a markedly rugged type. Many of the deep valleys of the Forest were cut in Triassic times along shatter belts or fault zones, and have now been re-exposed. The steepness of the original mountain slopes may be gauged by the great variation in thickness of the Keuper Marl within short distances. Locally slump structures are developed in the Keuper Marl adjacent to the Charnian rocks. Small-scale features of the old desert surface are revealed in the course of quarrying operations. Locally the Marl fills U-shaped gullies which resemble the wadis of modern deserts. At Mountsorrel the granodiorite against which the Marl is banked shows smoothed and terraced surfaces caused by sand-blast action. Raw (1934) considered that in part some of these features may have been produced by similar action during late Pleistocene times.

Parts of the Keuper Marl sequence are rich in evaporites. Anhydrite and gypsum are widely distributed but the major developments of rock salt are associated with the areas of maximum thickness of the Keuper Marl, which invariably coincide with the major depositional basins of the Permo-Triassic. There are very large amounts of rock salt in Cheshire and Shropshire alone—69 cubic miles (288 km$^3$) is thought to be a conservative estimate. Here, borehole evidence, notably from the Geological Survey's Wilkesley Borehole, has revealed that the salt beds are in two main groups called the Lower and Upper Keuper Saliferous Beds, which together have an aggregate thickness of over 2000 ft (610 m); of this thickness more than 80 per cent is salt. The two groups are separated by 1000 to 2000 ft (305–610 m) of mudstone, the Middle Keuper Marl, which contains a subordinate proportion of gypsum and anhydrite. Thinner beds of salt occur in Staffordshire and Worcestershire. Salt deposits were proved in boreholes at Chartley and at Bagot's Park near Uttoxeter and were formerly worked at Weston-on-Trent. Brine is pumped at Stoke Prior near Droitwich, at Stafford and in Cheshire, and rock salt is mined at Winsford (p. 106).

Beds of massive gypsum precipitated during the evaporation of a 'dead sea' occur at two horizons in the Middle Trent area, the lower bed at about 140 ft (43 m) and the upper bed at about 60 ft (18 m) respectively below the top of the Keuper. Strings and nodules of fibrous gypsum are common throughout the Keuper Marl.

The great thickness of evaporites reflects the aridity of the Keuper climate with its low rainfall and high rate of evaporation which exceeded the supply of water into the depositional basins. Dolomite and gypsum, the least soluble of the common dissolved minerals, are disseminated throughout the Keuper Marl. At times conditions were such that continuous precipitation of either gypsum or halite occurred over large areas. Following Ramsay and Hull, Sherlock (1948) considered that these salts were precipitated from the water of an arm of the sea, but clearly the drying up of such an embayment could have produced only a small fraction of the known deposits. Either the region was repeatedly flooded by sea water which was repeatedly dried up, or more likely, there was a restricted connexion with the open sea, along which flowed saline replacement waters into a region of high evaporation. At other times sea water was unable to enter the region, either because of the relative lowering of sea level or because the supply of fresh water was sufficient to maintain a lacustrine environment.

**Rhaetic**
At the end of Keuper times subsidence of the Triassic landmass led to a widespread marine transgression, which rapidly covered the whole of Central England except for small areas of older rocks which stood out as islands. The deposits laid down in this sea formed the Rhaetic.

The outcrop of the Rhaetic beds separates the Keuper and Liassic rocks, and forms a narrow band running north-westward across the Midland Plain; outliers have been preserved at Knowle, the Needwood Forest district, Barrow Hill, near Burton-upon-Trent and in the Wem–Audlem area (Prees Syncline). The total thickness of the formation is usually between 30 and 40 ft (9 to 12 m). It has been divided into the Lower Rhaetic (or Westbury Beds) and the Upper Rhaetic (or Cotham Beds).

*Lower Rhaetic* (*Westbury Beds*). The dark grey ('black') shaly beds of the lowest part of the Rhaetic reflect the establishment of marine conditions throughout Central England. There is a non-sequence at their base and the colour of the Tea Green Marl contrasts strongly with that of the overlying Rhaetic. Vertebrate remains occur at several horizons. Occasionally such a 'bone-bed' is developed at the base where it rests upon the uneven and bored surface of the Keuper.

The Lower Rhaetic is very uniform consisting of dark grey mudstones and shales with thin silt and argillaceous sandstone bands. In the Banbury district they are 5 to 37 ft (1·5 to 11·3 m) thick. They were well exposed at the Glen Parva Brickworks, south of Leicester, where they consist of 20 ft 9 in of 'black shales' with the characteristic lamellibranch *Rhaetavicula* [*Pteria*] *contorta* and occasional remains of the brittle star *Ophiolepis*. At the base there is a pyritous sandy conglomeratic band containing phosphatic nodules, coprolites and vertebrate remains: particularly common are fish scales, spines and teeth of fish such as *Acrodus, Ceratodus, Hybodus* and *Gyrolepis*. At Stanton on the Wolds, south-east of Nottingham, only very rare bones occur in the base of the Lower Rhaetic black shales, but slightly higher there is a one-inch band of sand containing coprolites, pebbles and vertebrate remains including *Ceratodus* similar to those recorded at Leicester. Near East Leake, the total thickness of the Rhaetic is at least 50 ft (15 m) of which about 31 ft (9·5 m) belong to the Westbury Beds. The sudden variations in thickness which occur between here and Grantham are probably related to the presence of major structures in the underlying Carboniferous.

The Lower Rhaetic is 31 ft 1 in (9·5 m) thick in the Wilkesley Borehole, in the Shropshire–Cheshire Basin, and consists of dark grey mudstones containing a few thin grey limestones and lenticles and thin beds of silt and fine-grained sandstone. Layers rich in fish remains occur at several horizons, but no 'basal bone bed' is developed. The fauna includes *Chlamys valoniensis, Protocardia rhaetica, Rhaetavicula contorta* and '*Natica*' *oppelii*.

The dark pyritic argillaceous beds of the Lower Rhaetic were deposited in shallow water, possibly under reducing but not stagnant conditions. Periods of current activity are reflected by the lenticles and thin beds of sand and silt, some of which contain rolled vertebrate remains.

*Upper Rhaetic (Cotham Beds)*. The Lower Rhaetic often passes imperceptibly into the Upper Rhaetic. The passage beds which may be up to 5 ft (1·5 m) thick consist of medium grey blocky mudstones. The overlying Upper Rhaetic consists of greenish grey calcareous silty mudstones with occasional thin limestone bands. Near Evesham they are $19\frac{1}{2}$ ft (5·9 m) thick and have been described as grey, greenish and yellowish 'shales'. South of Leamington the formation contains occasional brownish and olive grey bands. At Leicester the Upper Rhaetic is $10\frac{1}{2}$ ft (3·2 m) thick and in the Wilkesley Borehole this division consists of 13 ft 3 in (4 m) of greyish green calcareous mudstone and micaceous siltstones which are commonly laminated and finely cross-bedded. Slump structures are present throughout. The lower part becomes increasingly coarser grained and darker grey downward.

The fauna of the Cotham Beds is limited, the most common fossil being the conchostracan phyllopod *Euestheria*. These may occur alone or associated with insect remains and plant-rich laminae with the liverwort *Naiadita lanceolata*. The pale-coloured and calcareous beds of the Upper Rhaetic were probably deposited in a shallow brackish or freshwater environment into which marine incursions may have taken place; these may be indicated by the occurrence of *Eoguttulina liassica* and other foraminifera. Locally the Upper Rhaetic may contain thin olive-coloured beds. This suggests that the continental conditions of Keuper times may have returned for short intervals during the deposition of the Cotham Beds.

**Mineralization and intrusive igneous rocks**

The Permo-Triassic sandstones of Cheshire and parts of Shropshire have been affected by a very low-grade, but widespread, barytes-rich mineralization. Copper ore occurs at or near the base of the Keuper in the neighbourhood of Alderley Edge in northern Cheshire, and at a number of other places notably along the line of the Bickerton–Peckforton Fault in southern Cheshire and near Clive in Shropshire. The Triassic is penetrated near Swynnerton in North Staffordshire, and at Grinshill in Shropshire, by dykes of basic rock with a north-north-westerly trend. Their intrusion was connected with the widespread igneous activity of the Tertiary period (see *British Regional Geology*, Tertiary Volcanic Districts). The Swynnerton dyke forms a nearly vertical wall cutting the Keuper Marl; it attains a maximum thickness of 100 ft (30·5 m), but locally it splits into several thin dykes. The Grinshill porphyritic dolerite dyke penetrates both the Bunter and Keuper Series. It has been altered by barytes-rich mineralization which must have post-dated the Tertiary intrusion (Poole and Whiteman 1966).

# 9. Jurassic

Jurassic rocks form the surface over much of the southern and eastern parts of Central England. The area is part of the broad Jurassic outcrop which extends from the Dorset Coast to Yorkshire. Jurassic rocks formerly covered much of the ground lying west of the main outcrop, and may have extended into Wales, but if so, they have since been removed by denudation. Small remnants are left as outliers. The largest is the Prees Syncline situated near Wem, about seventy miles from the main outcrop. The System contains a great thickness of sediments showing so much lateral and vertical variation that their detailed correlation has provided one of the more complex tasks of English stratigraphical palaeontology.

Jurassic deposition in the Midlands was dominated by the presence in the south of the London Platform (Anglo-Belgian Island), whilst in the west were the Severn and Shropshire–Cheshire basins with the Welsh Massif beyond. To the north, the southern end of the Pennines probably formed a promontory dividing the depositional area into an eastern and a western zone. The north-eastern limit of this basin was controlled by the Market Weighton Axis; although the eastern limit of sedimentation is uncertain, it lay beyond the present region.

The major marine transgression initiated in late Triassic times continued with some intermissions well into late Jurassic times, probably reaching its maximum extent during the deposition of the Kimmeridge Clay. Post-Jurassic erosion has removed great thicknesses of Jurassic rocks from Central England, but, in general, sedimentation in Lower Jurassic times was thickest in the west and north away from the Palaeozoic ridge of the London Platform. The structural history of these areas may be regarded as one of persistent but intermittent downward movement, certain parts subsiding more rapidly than others. Within the basins, minor tectonic elements have also influenced the depositional history. For example, near Stratford-upon-Avon, on the southward projection of the Sedgley–Lickey Axis, an attenuated Rhaetic sequence is overlain by Lower Lias with a basal conglomerate containing derived Rhaetic fossils, and north of the Melton Mowbray Anticline, where the lowest beds of the Lower Lias show a marked increase in thickness. The area of maximum deposition varied; thus in Upper Lias times it is thought to have migrated southwards from Lincolnshire into the Midlands. In the extreme south-east of the region near the London Platform, the Lower and Middle Jurassic sediments show variations in thickness as a result of periods of overlap and overstep contrasting with periods of regression or offlap.

The earliest Jurassic strata, the Lias, are predominantly fine-grained argillaceous and calcareous rocks deposited beneath the waters of the transgressive sea which first invaded much of Britain at the end of Keuper times. During the Middle Lias a change occurred, possibly a shallowing of the sea, which permitted the deposition of silts and thin calcareous sandstones and later produced a shallow-water environment in which ferruginous and calcareous shelly sediments were laid down to form the Marlstone Rock Bed. Upper Lias times saw the return of dominantly argillaceous deposition. During the Middle Jurassic the environmental conditions varied greatly, the shallow sea being converted at times into deltaic or estuarine areas where thin plant beds or seatearths were formed at periods of relatively low sea level. The marine deposits laid down at this time include calcareous sediments represented by the Northampton Sand, the Lincolnshire Limestone, the Great Oolite Limestone (Blisworth Limestone) and the Cornbrash. The Upper Estuarine Limestone and parts of the Great Oolite Clay (Blisworth Clay) were formed in more restricted marine environments, whilst the Estuarine Series includes deltaic or coastal plain deposits. At the end of Cornbrash times it is possible that slight uplift of the land led to the introduction of fine sand and silt (Kellaways Beds), but soon the deposition of the Oxford Clay marked a return to the conditions of argillaceous sedimentation very similar to those which existed during Liassic times.

Many of the Jurassic rocks are richly fossiliferous. Amongst the most significant fossils are the ammonites some of which show rapid evolutionary changes, the newly developed forms replacing existing species to give a sequence of faunas; on this basis a table of characteristic ammonite zones has been established. Some characteristic Jurassic fossils are shown on Plate IX.

## Lias

The term Lias (Lyas), first used in geological work by John Strachey in 1719, is an old West of England quarryman's term, and was applied primarily to thin beds of muddy and shelly limestone which occur in the lower part of the Lower Lias (and also in the Purbeck rocks of Dorset). Buckland (*in* Phillips 1818) drastically extended the original usage and the term is now applied to the predominantly argillaceous formations of the Lower Jurassic.

THE AMMONITE ZONES OF THE LIAS

| | |
|---|---|
| Upper Lias | *Dumortieria levesquei*<br>*Grammoceras thouarsense*<br>*Haugia variabilis*<br>*Hildoceras bifrons*<br>*Harpoceras falciferum*<br>*Dactylioceras tenuicostatum* |
| Middle Lias | *Pleuroceras spinatum*<br>*Amaltheus margaritatus*<br>*Prodactylioceras davoei*<br>*Tragophylloceras ibex*<br>*Uptonia jamesoni* |
| Lower Lias | *Echioceras raricostatum*<br>*Oxynoticeras oxynotum*<br>*Asteroceras obtusum*<br>*Caenisites turneri*<br>*Arnioceras semicostatum*<br>*Arietites bucklandi*<br>*Schlotheimia angulata*<br>*Alsatites liasicus*<br>*Psiloceras planorbis* |

**Lower Lias**

The Lower Lias crops out in the Vale of Evesham and continues through Rugby, Leicester and Melton Mowbray to the north-eastern corner of the region and thence into Lincolnshire. Tongues extend northward to Droitwich and Henley-in-Arden. A small outlier occurs at Knowle and a second at Prees on the borders of Shropshire and Cheshire. Three lithological divisions have been recognized, the White Lias, the Blue Lias and the Lower Lias Clay. These are formations and not biostratigraphic units, and their constituent beds may transgress major biostratigraphic junctions.

Fossils are generally abundant in the Lower Lias and include representatives of many animal groups. The wide variety of ammonites found is particularly useful for the accurate correlation of horizons in a fairly uniform argillaceous sequence. Brachiopods, e.g. *Calcirhynchia calcaria*, are locally abundant whilst lamellibranchs such as *Gryphaea* (Pl. IX, fig. 1), *Liostrea* and *Modiolus* occur at many levels. Corals, echinoderms, crustacea and vertebrates are also found. The Lower Lias was deposited in a quiet-water marine environment, the abundance of trace fossils of burrowing habit suggesting that despite its present pyritic character, stagnant conditions existed only rarely. During the accumulation of the Blue Lias sediments, deposition of calcium carbonate, either primary or diagenetic, was important. Periods of increased current activity possibly associated with shallowing are represented by thin skeletal marls and limestones or, near the top of the sequence, clays with shell debris, phosphatic nodules and glauconite.

*White Lias*. The White Lias is by original definition part of the Lias and is therefore traditionally included in the Jurassic. In character, however, these pale marls and limestones resemble the top of the underlying Cotham Beds, the junction of the two formations being transitional in Central England. The White Lias crops out in the area south of Stratford-upon-Avon and can then be traced northward to Wigston south of Leicester, beyond which the facies is absent, though at Costock, south of Nottingham, thin marls and limestones with *Modiolus* cf. *langportensis* may be of similar age. The formation consists of beds of white or pale grey, fine-grained, commonly porcellanous limestones (calcite mudstones) with thin clay partings. Locally the limestones are very pure and were probably chemically precipitated in a shallow marine environment. Individual limestone bands may be penetrated by boring organisms, whilst others are pebbly and occasionally show evidence of penecontemporaneous brecciation. Near Evesham the White Lias is absent. In Warwickshire the formation is overlain non-sequentially by the Lower Lias, the basal bed of which is a shelly conglomerate containing a mixed Lower Lias and derived Rhaetic fauna. Near Rugby the basal bed of the Lower Lias Clay is highly bituminous and pyritic and infills hollows in the topmost bed of the White Lias.

The White Lias fauna contains a variety of lamellibranchs including *Dimyopsis intusstriata*, *Lima valoniensis*, *Liostrea hisingeri*, *Meleagrinella fallax*, *Modiolus langportensis* and *M. minimus;* foraminifera, echinoderms, gastropods and ostracods also occur.

*Blue Lias and Lower Lias Clay*. The Blue Lias consists of an alternation of fine-grained argillaceous limestones (Hydraulic Limestones) with grey calcite mudstones and shales. They have been extensively dug for the manufacture of lime and cement. The limestones are usually confined to some horizons within the lower zones of the Lower Lias; the rest of the formation consisting of bluish grey slightly calcareous mudstones and shales is very poorly exposed.

A borehole at Mickleton eight miles east of Evesham proved 961 ft (293 m) and included both White and Blue Lias beds. Other boreholes show that the maximum thickness of the Lower Lias near Stow on the Wold may reach 600 ft (183 m) and around Rugby it reaches between 650 and 750 ft (198–229 m). Traced south and south-eastward from the outcrop in the direction of the London landmass the formation is attenuated and is finally overlapped or overstepped by the Upper Lias and the Middle Jurassic formations.

Near Binton, west of Stratford-upon-Avon, the basal Lower Lias is a pebbly shelly limestone which includes fossils derived from the Rhaetic. To the east up to 40 ft (12 m) of argillaceous strata separate the White and Blue Lias; the uppermost beds of this clay contain ammonites characteristic of the *Schlotheimia angulata* Zone, whilst the overlying Blue Lias limestones lie within the *S. angulata* and the lower part of the *Arietites bucklandi* Zone. At Victoria Quarry, Rugby, the Blue Lias is about 75 ft (23 m) thick and consists of at least 35 beds of limestone alternating with bluish grey shale, ranging from the *Psiloceras planorbis* Zone up to the *Arietites bucklandi* Zone. In north Leicestershire and Nottinghamshire the Blue Lias limestones are about 25 ft (7·6 m) thick and lie within the *P. planorbis* and *S. angulata* zones. The underlying pre-*planorbis* Beds consist of fine-grained calcite-mudstones and shales which may contain an abundance of saurian remains associated with a fauna of marine foraminifera, lamellibranchs and ostracods, but ammonites are not found. Near Grantham the Blue Lias, including the pre-*planorbis* Beds, is up to 30 ft (9·1 m) thick and rests directly on the Rhaetic. It is overlain by clays of the *S. angulata* Zone. The higher beds of the Lower Lias are predominantly clay but include beds of ferruginous limestone, iron-rich nodules and sandy clay similar to those which characterize the Lower Lias farther south.

Lower Lias outliers are known in the Shropshire–Cheshire Basin. The largest, the Wem–Audlem outlier, is a faulted basin with a length of 14½ miles (23·3 km) and a maximum width of about 5 miles (8 km). The outcrop is largely obscured by drift but a borehole at Wilkesley proved 458 ft (140 m) of Lower Lias including all the zones up to the Zone of *Arnioceras semicostatum*. It is possible that the total thickness of the Lower Lias within this area may be as much as 1000 ft (305 m).

**Middle Lias**

The Middle Lias is divided into a lower group consisting predominantly of silts and clays (Zone of *Amaltheus margaritatus*) and an upper calcareous and ferruginous unit, the Marlstone Rock Bed (Zone of *Pleuroceras spinatum*). The outcrop can be traced throughout the region from Edge Hill, near Banbury, through Daventry and Oakham to Grantham. The Rock Bed produces a marked escarpment, as at Edge Hill, whilst the underlying silty beds give rise to a light soil which is frequently covered with gorse thickets. The Middle Lias crops out in the Wem–Audlem outlier, where micaceous sandy clays and sandy 'marlstones' of the *A. margaritatus* Zone are overlain by 8 to 10 ft of Marlstone Rock Bed with *Pleuroceras spinatum*.

At the close of Lower Liassic times slightly coarser material was swept into the Liassic sea, and the micaceous silts and clays of the *A. margaritatus* Zone were accumulated. Progressive shallowing of the sea was associated with the deposition of the silty and sandy ferruginous and occasionally false-bedded limestones which locally occur in the upper part of the Zone. The widespread basal conglomerate of the Rock Bed indicates a period of intensive current sorting. The overlying beds are commonly cross-stratified and contain an abundance of shell debris commonly associated with ooliths which suggests deposition in a relatively shallow current- or wave-swept sea. The proportion of ferruginous material and the iron content of the Rock Bed varies throughout its outcrop. Generally the most ferruginous beds are almost sand free and suggest that local conditions favoured chemical precipitation, whilst elsewhere, possibly in marginal zones (sediment traps), a large amount of sand was washed into the sea.

*Middle Lias Clays and Silts.* In the Bredon Hill–Banbury district the lower part consists of silty clays with thick bands of ferruginous concretions whilst near the top ferruginous sandstones and sandy limestones are developed. At Edge Hill the total thickness exceeds 130 ft (39·6 m); near Banbury it is reduced to 30 ft (9·1 m), but it increases again farther north to attain a maximum of 120 ft (36·6 m) at Grantham. In the Rugby area the Middle Lias consists largely of silts and silty clays with thin ironstone bands near the top. North of Market Harborough it becomes increasingly argillaceous, but at Melton Mowbray a thin sandy micaceous ironstone is present at the base. Amaltheid ammonites are fairly common throughout, and are associated with lamellibranchs, e.g. *Modiolus scalprum*, *Oxytoma inequivalve*, *Pronoella intermedia* and *Protocardia truncata*.

*Marlstone Rock Bed.* This formation was sufficiently ferruginous to be exploited as a low-grade iron ore in the Banbury district and in North Leicestershire and South Lincolnshire (see Fig. 16). In Oxfordshire the ironstone is best developed in the Wroxton–Bloxham area and around Kings Sutton and Adderbury and has been worked in the Charwelton–Byfield area of Northamptonshire. At Edge Hill the Rock Bed is at least 25 ft (7·6 m) thick but to the south-west and to the east it is thinner and is represented by ferruginous sandy limestones and calcareous sandstones. The ore consists of calcitic sideritic chamosite oolite and sideritic shell-fragmental limestone. At the base there is a distinctive conglomerate containing pebbles of chamosite and phosphatic mudstone and reworked fossils, particularly belemnites, which rests non-sequentially on the clays of the *A. margaritatus* Zone. This horizon is persistent and in some areas it is the only representative of the Rock Bed. The northern limit of the workable ironstone is near Hellidon, south of Daventry. Farther north the Rock Bed is thin and conglomeratic or locally absent.

Around Market Harborough the Marlstone Rock Bed is absent but reappears farther north and increases in thickness until at Tilton, Leicestershire, it attains its maximum development in the Midlands, being 30 ft (9·1 m) thick. The upper part is a workable ironstone; the lower division, 'the sandrock', is a hard calcareous sandstone with a conglomeratic bed at the base. Around Oakham the Rock Bed is up to 10 ft (3 m) thick and consists of ferruginous and sandy limestones. North of Melton Mowbray the thickness varies from 8 to 30 ft (2·4–9·1 m); the upper part is a workable ironstone and has been extensively exploited.

Ammonites are extremely rare in the limestone and sandstone facies of the *P. spinatum* Zone which are developed in the Midlands. 'Nests' of brachiopods abound and include *Tetrarhynchia tetrahedra* (Pl. IX, fig. 2), *Gibbirhynchia micra*, *G. northamptonensis* and *Lobothyris punctata;* whilst lamellibranchs present include *Entolium liasianum* and *Pseudopecten aequivalvis*.

## Upper Lias

The Upper Lias crops out over a wide area west of Brixworth and around Uppingham; elsewhere the outcrops are narrow and discontinuous being restricted to the sides and floors of valleys which have been cut through the platform produced by the outcrop of the more resistant Middle Jurassic sediments.

The main mass of the Upper Lias of the Midlands is of comparable age to the Cotteswold Sand and Upper Lias clay of the South Cotswolds. The maximum thickness of Upper Lias occurs in Northamptonshire (250 ft; 76·2 m). Borehole evidence shows that the thickness decreases south-eastwards towards the London landmass which continued to be a positive area. Study of the ammonite zones in the region has shown that there is a progressive thickening of the higher zones at the expense of the lower zones, when traced from north to south.

In Northamptonshire the beds below the *Peronoceras fibulatum* Subzone of the *Hildoceras bifrons* Zone contain thin argillaceous limestones some of which contain pseudo-ooliths. These and the interbedded clays contain an abundance of ammonites. Near the base of the Upper Lias, beds of hard fissile limestone interbedded with paper shales, all with fish debris, are developed. They rest on a very fossiliferous 'Transition Bed' which has a limited geographical distribution. It was once considered to represent a zone or subzone characterized by *Tiltoniceras acutum* but this bed is now included in an undivided *Dactylioceras tenuicostatum* Zone. After this period of intermittent carbonate deposition there was a return to quiet water conditions similar to those of Lower Lias times, the remainder of the Upper Lias consisting of bluish grey mudstones with calcareous and ferruginous nodules.

The fauna of the Upper Lias is rich in cephalopods and lamellibranchs. The ammonites include many species of *Dactylioceras* (Pl. IX, fig. 3), *Harpoceras* and *Hildoceras;* belemnites are common as are the lamellibranchs *Inoceramus dubius*, *Nuculana ovum* and *Steinmannia bronni*.

## Inferior Oolite Series

Clays, silts, ferruginous sandstones and oolitic limestones are some of the rock types included in the Inferior Oolite Series. They contrast strongly with the uniform argillaceous rocks of the underlying formation upon which they rest unconformably. Around Northampton the Northampton Sand rests upon mudstones of the *Haugia variabilis* Zone: traced northward these latter beds are overstepped and north of Grantham the Inferior Oolite comes down on to the *Peronoceras fibulatum* Subzone of the *H. bifrons* Zone. The presence of derived Upper Lias limestone pebbles and fossils in the base of the Northampton Sand at some localities shows that the surface of the Lias underwent erosion prior to the deposition of the Sand. The Northampton Sand appears to have been laid down in a shallow sea, in parts of which the chemical precipitation of iron compounds predominated. Subsequent further shallowing of the sea permitted the formation of a series of deltas or estuarine strand-flats on which the sediments of the Lower Estuarine Series accumulated. Later a marine transgression introduced conditions favourable to the formation of the Lincolnshire Limestone. In Rutland carbonate deposition was continuous but to the south the lower part of the formation was deeply channelled before the deposition of the upper beds (see Fig. 15).

The instability of the region during the deposition of the Inferior Oolite Series is reflected by the presence of numerous non-sequences and minor unconformities. Owing to the lack of ammonites detailed correlation of the Series on a zonal basis is incomplete. The general succession in the region is shown in Figure 15.

*Northampton Sand.* The type locality of the Northampton Sand is north and west of Northampton, where the formation attains a maximum thickness of nearly 70 ft (21·3 m). The basal part is developed as an ironstone which is overlain by massive yellow or brown sandstone with beds of fissile calcareous sandstone. These sandstones (sometimes known as the Variable Beds) are absent over most of the 'Ironstone Field' where the formation is commonly only 15–25 ft (4·6–7·6 m) thick. Layers of clay may be interbedded with the lowest horizons. The Northampton Sand extends northward into Lincolnshire, but thins markedly to the south-east of Northampton, and dies out south of Towcester. The formation persists into the Banbury–Brackley district where it consists of calcareous sandstones and sandy limestones, some of which are ferruginous. This facies is intermediate between the typical Northampton Sand and the *scissum* Beds of the Cotswolds. In the ironstone facies the Northampton Sand consists of chamositic, kaolinitic, sideritic and limonitic oolites, with shelly oolites, sideritic and chamositic mudstones. The nature and proportion of the allogenic grains including sand, shells and shell debris vary in the different rock types. Hollingworth and Taylor (1951) have proposed a lithological classification the sequence of which can be recognized over small areas. Each litho-facies is thought to represent a slight variation within a single depositional environment.

The fauna of the Northampton Sand consists mainly of lamellibranchs but owing to extensive decalcification most of the fossils are preserved as moulds. The few ammonites which have been found include *Bredyia newtoni, Leioceras opalinum, Lytoceras aalenianum* and *Tmetoceras scissum.* These are indicative of the upper part of the *L. opalinum* Zone. Lamellibranchs are locally abundant and include *Astarte (Neocrassina) elegans, Ceratomya bajociana, Gervillia northamptonensis, Myophorella (Vaugonia) v-costata* (Pl. IX, fig. 4), *Variamussium pumilum* and *Lima (Plagiostoma)*

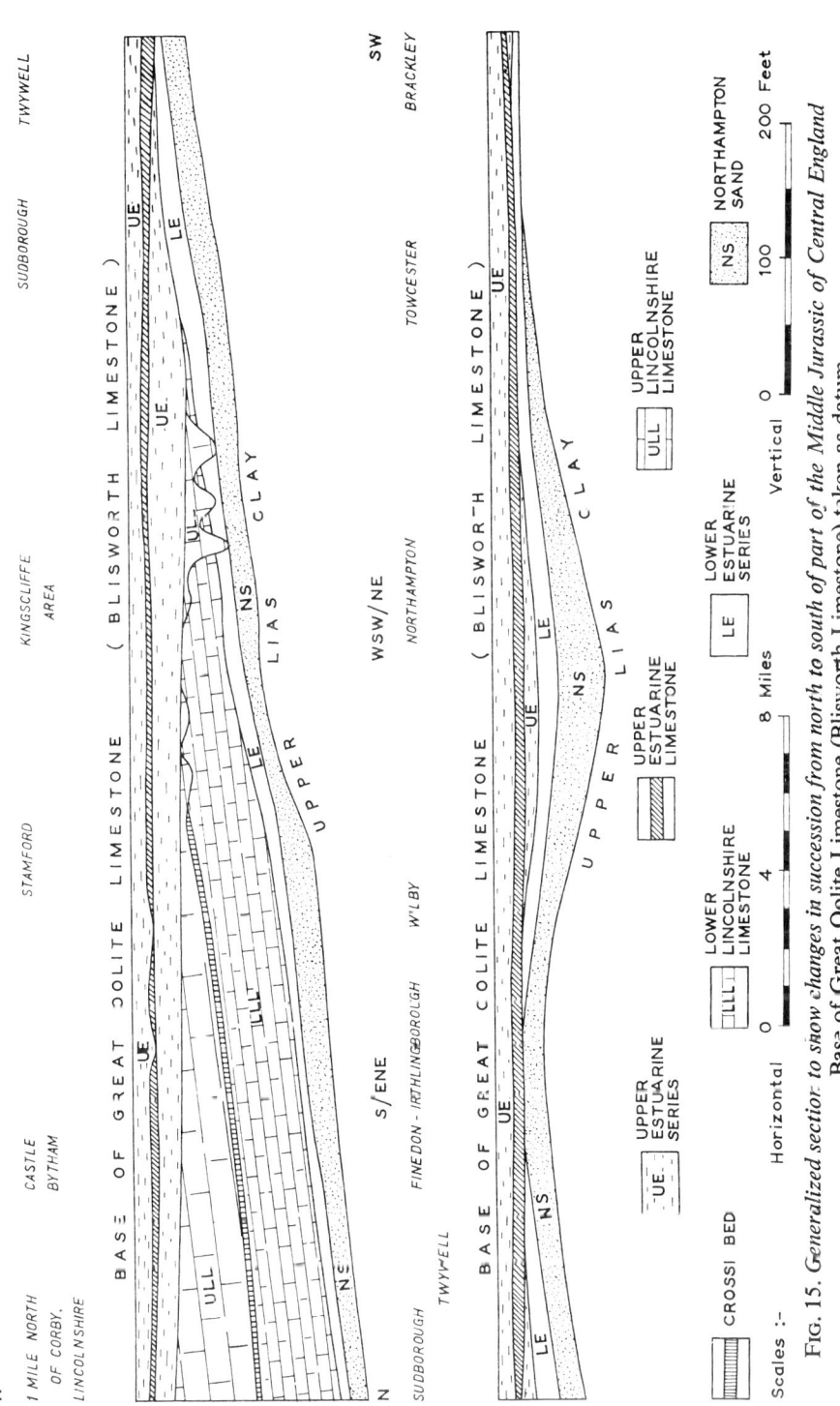

Fig. 15. *Generalized section to show changes in succession from north to south of part of the Middle Jurassic of Central England. Base of Great Oolite Limestone (Blisworth Limestone) taken as datum*

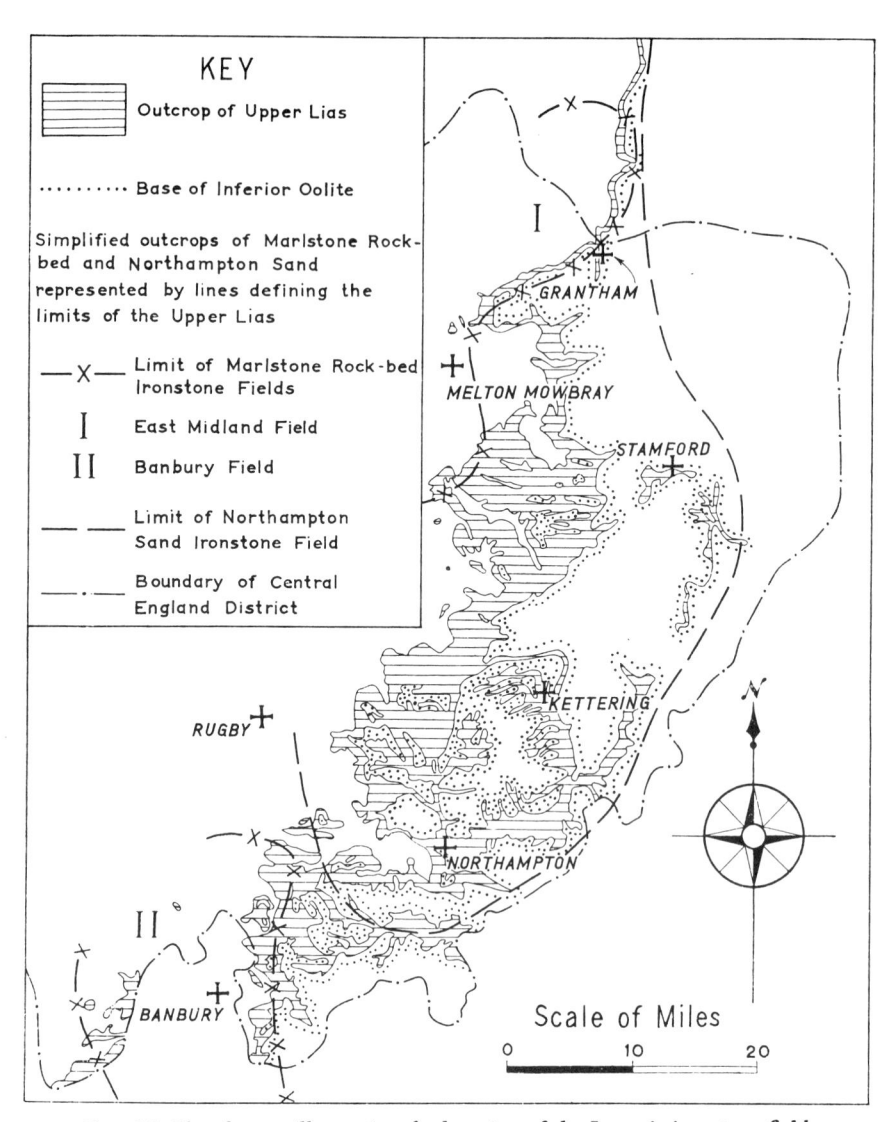

FIG. 16. *Sketch-map illustrating the location of the Jurassic ironstone fields and limits of workable ironstone*

*spp.;* gastropods such as *Nerinella cingenda* and brachiopods of the genera *Lobothyris*, *Homoeorhynchia* and *Tetrarhynchia* are also present.

Taylor (1949, p. 79) suggested that the Northampton Sand was deposited in 'a shallow epi-continental sea, gulf or lagoon freely connected with the open sea and subjected to considerable wave and current action'. Evidence of contemporaneous erosion occurs throughout the formation. Sandy marginal facies are developed around Wellingborough and Northampton, between Oakham and Grantham and to the north and south of Stamford (see Fig. 16).

*Lower Estuarine Series.* The formation has been recognized throughout Central England from Grantham south and south-westwards to Banbury. The maximum thickness is about 25 ft (7·6 m). It consists of pale grey and white sands, grey to lilac tinted silts and clays with occasional calcareous sandstones and carbonaceous clay. It normally overlies the Northampton Sand. There is a major non-sequence or unconformity at the base, the formation locally overstepping the Northampton Sand and coming to rest on Upper Lias. Near Grantham the Lower Estuarine Series fills elongated channel-like depressions similar to washouts which are incised through the Northampton Sand and into the top of the Upper Lias Clay. North of Kettering it has been cut out completely by channels infilled with Lincolnshire Limestone (Fig. 15). Farther south it is also locally absent, the Upper Estuarine Series resting directly on the Northampton Sand or Upper Lias.

Lithologically two facies can be recognized. Around Corby, Northamptonshire, a predominantly argillaceous division separates two groups of sandy and silty beds. Usually at the top of the middle division is the 'Coaly Bed' a black carbonaceous shale which rests on a band of sandy silt with vertical rootlets. To the south and east of Kettering the formation consists largely of white sands and silts, which attain their maximum thickness in a trough running north-east from Towcester to Oundle. Coaly beds are again developed near Stony Stratford and south-west of Brackley.

The organic remains are almost entirely plant fragments, though ostracods are occasionally found. Thin marine beds with gastropods, *'Cyrena' sp.* and *Corbula sp.* are recorded from Colsterworth, 7 miles (11 km) south of Grantham. Lithologically many beds resemble those recorded from the Coal Measures, ganisters and fireclays being present. Locally the underlying Northampton Sand has been leached and in places sphaerosiderite has been formed beneath pockets infilled with Lower Estuarine sand. Small-scale channelling occurs in the upper part of the formation. Taylor (1963) suggested that the formation was deposited in a low-lying coastal plain or delta-flat environment.

*Lincolnshire Limestone.* The maximum recorded thickness of the Lincolnshire Limestone, 132 ft (40·2 m), is from a well at Boothby Pagnell, south-east of Grantham. At the outcrop, several miles to the west, the thickness is 60 ft (18·3 m). Where unobscured by drift it forms a distinctive limestone plateau. Traced southward the thickness decreases and the formation disappears in the Kettering–Oundle district.

Two divisions, an upper and a lower, are recognizable. In the north, between Grantham and Stamford, a bed of limestone characterized by the spinose rhynchonellid *Acanthothiris crossi* (Pl. IX, fig. 5) marks the top of the Lower Lincolnshire Limestone and forms a readily mappable horizon. Here the Upper and Lower Lincolnshire Limestone are generally conformable, although locally the *crossi* Bed is cut out by a small intraformational unconformity. South of Stamford the *crossi* Bed and overlying Upper Lincolnshire Limestone are usually absent beneath the unconformity at the base of the Upper Estuarine Series. The Upper Lincolnshire Limestone is present however near Barnack and also in the Kingscliffe district north of Kettering where it rests on the channelled and eroded surface of the Lower Lincolnshire Limestone, Lower Estuarine Series and exceptionally on the Northampton Sand.

South of Grantham the Lower Lincolnshire Limestone consists of siliceous slightly ferruginous limestones, overlain by cream or white oolitic and pisolitic limestones and oolites, which are succeeded by the white limestones and marls of the *crossi* Bed. Around Stamford the basal beds of the Lincolnshire Limestone are sandy and locally the Collyweston Slate is developed. It appears to be conformable on the Lower Estuarine Series and consists of fissile sandy limestones which yield specimens of *Gervillella acuta, Modiolus (Inoperna) plicatus, Pinna sp.* and plant debris. The rest of the Lower Lincolnshire Limestone includes beds of oolith-pellet limestone, oolitic limestones and shell-fragmental limestones.

North of Stamford the Upper Lincolnshire Limestone is divisible into two groups, separated by a minor unconformity: the lower group includes the freestones of Ketton and Stamford, whilst the upper consists of coarse shelly oolites and pisolites which fill hollows and channels eroded in the surface of the older rocks. In the Kettering district, the upper surface of the Lower Lincolnshire Limestone is extensively bored when overlain by the Upper Lincolnshire Limestone. The latter is similar lithologically to the lower division but is usually of coarser texture and contains an abundance of shell debris. The lowest beds which fill the channel-like depression north-east of Kettering are strongly current-bedded coarse shelly oolites. Taylor (1963) has suggested that these channels were formed by submarine erosion. The higher beds include oolites such as the Weldon Freestone and lenticular oyster-rich oolites such as the 'rags' of Barnack and Weldon.

In Lincolnshire ammonites of the *Sonninia sowerbii* Zone including *Hyperlioceras* aff. *discites* occur in beds below the *crossi* horizon. The Lower Lincolnshire Limestone is highly fossiliferous, and yields many molluscs including *Pholadomya fidicula, Gervillella acuta* and *Bactroptyxis cotteswoldiae*. The Upper Lincolnshire Limestone contains a rich and varied fauna. Brachiopods abound and include *Microrhynchia barnackensis, Weldonithyris weldonensis* and *Zeilleria wilsfordensis*. Corals such as *Montlivaltia stutchburyi* are associated with species of *Nerinea, Anisocardia, Astarte, Chlamys* and *Liostrea*.

The Collyweston Slate and associated basal deposits with their high proportion of terrigenous detritus were formed during the initial stages of a marine transgression. In Lincolnshire these beds are represented by soft, yellow, micaceous sandstones more akin to the Lower Estuarine Series than the Lincolnshire Limestone. The remainder of the Lincolnshire Limestone was deposited in relatively warm shallow current-agitated water which supported a large and varied fauna.

Plate IX  Characteristic Jurassic Fossils

1. *Gryphaea arcuata* Lamarck, side view—Lower Lias. 2a and b. *Tetrarhynchia tetrahedra* (J. Sowerby), dorsal and anterior views—Middle Lias. 3. *Dactylioceras directum* (S. S. Buckman), side view—Upper Lias. 4. *Myophorella v-costata* (Lycett), side view of cast—Northampton Sand. 5a and b. *Acanthothiris crossi* (J. F. Walker), dorsal and anterior views—Lower Lincolnshire Limestone. 6. *Nucleolites clunicularis* (Wm. Smith), dorsal view—Great Oolite Series. 7a and b. *Epithyris oxonica* Arkell, dorsal and anterior views—Great Oolite Series. 8. *Modiolus imbricatus* (J. Sowerby), side view—Great Oolite Series. 9. *Cylindroteuthis puzosiana* (d'Orbigny), Oxford Clay.

Plate X

A. Upper Lincolnshire Limestone (Weldon Stone), Great Weldon (A.8353)

(*For full explanation see p. ix*)

B. Glacial Deposits, Shawell Gravel Pit, near Rugby (A.10128)

## Great Oolite Series and Cornbrash

The beds constituting the 'Great Oolite Series' embrace in upward sequence, the Upper Estuarine Series, the Great Oolite Limestone (Blisworth Limestone) and the Great Oolite Clay (Blisworth Clay). The latter passes laterally into the Forest Marble at Lillingstone Lovell, north of Buckingham and again south-west of Brackley. In Central England the Great Oolite Series is almost confined to Northamptonshire and parts of Rutland. During the closing phases of the deposition of the Inferior Oolite Series elevation of the sea floor occurred. The minimum uplift was in the north around Grantham, where the Upper Estuarine Series rests with slight unconformity upon the Upper Lincolnshire Limestone. In south Northamptonshire, where the Lincolnshire Limestone is thin and impersistent, the transgressive Upper Estuarine Series oversteps the Lincolnshire Limestone and rests locally on Lower Estuarine beds, Northampton Sand or Upper Lias.

The Upper Estuarine Series was probably laid down in coastal swamps or marshes subject to intermittent marine incursions which preceded the extension of the transgressive sea in which the overlying Great Oolite Limestone was accumulated. During the deposition of the Limestone strong currents swept the shell debris into banks, elsewhere ooliths were formed. At times tranquil conditions allowed the deposition of thin clay bands and the formation of calcite-mudstone, largely by chemical precipitation. In most of Central England, the development of coastal lagoons ended the deposition of the Great Oolite Limestone. Periodic incursions of the sea occurred throughout Great Oolite Clay times. In the extreme south marine conditions were more persistent and in this area the shelly limestones and silty clays of the Forest Marble were laid down.

*Upper Estuarine Series.* The Series consists of light and dark green and grey clays and silts, with occasional bands of black and buff hue, associated with thin limestones and marls. The formation attains a maximum thickness of 40–45 ft (12·3–13·4 m) in the Stamford–Oundle area, whilst south of Kettering it is reduced to 8 ft (2·4 m) locally but increases to about 30 ft (9·1 m) in the vicinity of Towcester.

In Rutland and Lincolnshire, where the Upper Estuarine Series rests directly on the Lincolnshire Limestone, a distinctive band of nodular ironstone is present at the base. Generally the junction beds are conformable but south of Kettering the Upper Estuarine Series overlies the Lower Estuarine Series, both the Lincolnshire Limestone and the ironstone junction bed being absent. Near Northampton the upper formation is channelled into the lower and includes a basal ironstone nodule band which, in places, appears to be a bedded deposit but elsewhere is of secondary origin.

North of Oundle two marine horizons occur in the lower part of the Upper Estuarine Series. The lowest bed may be equivalent to the marine limestones and marls (Upper Estuarine Limestone) which are developed farther south. In south Northamptonshire the lower group of Upper Estuarine clays and silts is absent and the Upper Estuarine Limestone rests directly on older strata. The limestones are usually of shell-fragmental, argillaceous or marly character but they may also be hard, massive, and non-fossiliferous as near Fotheringhay, and are sometimes interbedded with clays and marls containing *Liostrea hebridica* and *Modiolus imbricatus* (Pl. IX, fig. 8). Near Towcester false-bedded sandy limestones are developed and are associated with sandy oyster-rich limestones and oyster lumachelles. Other fossils common in the limestones are *Kallirhynchia sp.*, *Eomiodon fimbriatus*, *Placunopsis socialis* and *Protocardia citrinoidea*, whilst *Lingula kestevenensis* and *Bakevellia waltoni* have been recorded from the clays below the Limestone and *Corbula buckmani* and *Tancredia* cf. *similis* are present in the clays above.

The Lower and Upper Estuarine Series show lithological similarities; both contain rootlet beds but the sediments of the latter are generally finer grained and contain more marine intercalations. Taylor (1963) has suggested that the formation was deposited in a marsh environment on a low-lying coastal plain or delta, the most important marine invasion being now represented by the beds of the Estuarine Limestone. The appearance of *Kallirhynchia sharpi* in the topmost clays of the Upper Estuarine Series heralded the establishment of the sea in which the predominantly carbonate deposition of the Great Oolite Limestone occurred.

*Great Oolite Limestone* (*Blisworth Limestone*). The formation ranges from 15–25 ft (4·6 to 7·6 m) in thickness and comprises a variable series of cream-weathering calcite-mudstones, calcite-siltstones, shelly limestones, shell-fragmental limestones, pseudo-oolitic and oolitic limestones, some of which are freestones. Most exposures show an alternation of these lithologies. Thin clay bands are developed locally.

The most common fossils are oysters, numerous oyster lumachelles being present. The Great Oolite Limestone contains an abundant fauna throughout Central England but traced northward this becomes progressively impoverished as the formation thins, until the Limestone dies out south of the Humber. In the Kettering district, the *sharpi* Bed, a white rubbly argillaceous limestone with interbedded bands of clay, is crowded with *Kallirhynchia sharpi* and forms a distinctive marker band at the base of the formation. Other brachiopods present include *Epithyris oxonica* (Pl. IX, fig. 7) and various species of *Kallirhynchia* and *Digonella*. Lamellibranchs are particularly numerous and include *Astarte* (*Neocrassina*) *rotunda*, *Liostrea hebridica* and *Parallelodon hirsonensis*. At Orton, an interesting series of fish remains has been found, including complete palates of such forms as *Macromesodon bucklandi*.

Taylor (1963) suggests that the Great Oolite Limestone in the Kettering district with its abundance of shell-fragmental limestones and scarcity of ooliths indicates a depositional environment similar to that which exists today in the marginal areas of the Bahama Banks. In the Towcester district the Limestone contains an important proportion of oolites and is lithologically similar to the Great Oolite Limestone of the North Cotswolds, which may have been deposited by a series of laterally migrating creeks in the intertidal zone of an oolitic limestone forming sea (Klein 1963).

*Great Oolite Clay* (*Blisworth Clay*). The formation consists largely of vividly coloured clays, ranging in shade from inky-blue to black, dark grey, green and purple. The clay is plastic and tenacious, and locally a ferruginous clay ironstone band is developed at the base. The clay beds are almost completely devoid of fossils but thin beds of marl which yield *Liostrea hebridica* and *L. subrugulosa* are present in some areas. The thickness varies from 10 to 25 ft (3–7·6 m). Around Towcester a bed of shell-fragmental limestone occurs in the lower part of the formation whilst to the southeast a laminated limestone with plant fragments occurs at about the same horizon. Near Buckingham the Great Oolite Clay passes laterally into the Forest Marble, which consists of pale grey silty clay with thin silty sandstone ribs and micro-cross-laminated sandy limestones. Coarse, blue, oyster-rich limestones are found at the base.

The wide distribution, the fine-grained nature of the sediments and the uniformity of the Great Oolite Clay facies in the East Midlands suggests that these deposits were accumulated in an enclosed lagoonal basin. Only rarely did the sea invade this area, but in the extreme south of the region, the development of the Forest Marble facies indicates the establishment of a shallow sea or tidal zone, into which a large amount of fine-grained material and shell debris was transported.

*Cornbrash.* The Cornbrash consists of distinctive, reddish brown weathering, shelly limestones; locally it includes marls and fine-grained limestones and beds containing a varying proportion of shell debris and sand. The thickness varies from a nodular band a few inches thick to about 10 ft (0·05–3 m). When drift free, the outcrop produces a prominent topographic feature.

The formation may be classified as follows:

|  | Characteristic brachiopods | Ammonite zones |
|---|---|---|
| Upper Cornbrash | 4. *Microthyridina lagenalis*<br>3. *Microthyridina siddingtonensis* | *Macrocephalites macrocephalus* |
| Lower Cornbrash | 2. *Obovothyris obovata*<br>1. *Cererithyris intermedia* | *Clydoniceras discus* |

There is a distinct lithological and palaeontological break between the two divisions and this line of unconformity has been used as the Middle Upper Jurassic junction. The thickness variation is partly due to the preferential development or absence of either the Lower or the Upper Cornbrash and is partly the result of penecontemporaneous erosion, beds containing pebbles of Cornbrash-type limestone occurring at several horizons within the formation.

Although only a few feet thick, the rock has been much dug for local roadstone in the past, but it is now but little quarried and exposures are correspondingly rare. With the Great Oolite Clay, the Cornbrash forms the overburden in quarries for Great Oolite Limestone at Thrapston.

## Kellaways Beds and Oxford Clay

The Kellaways Beds consist of clays and sands: locally they may be divided into a basal Kellaways Clay and the Kellaways Sand, the latter when massive and cemented being known as the Kellaways Rock. The maximum thickness of the Kellaways Beds and Oxford Clay is probably about 300 ft (92 m), but in this region the upper part has been removed by recent denudation. The ammonite zones of these beds are in upward sequence, *Sigaloceras calloviense, Kosmoceras jason, Erymnoceras coronatum, Peltoceras athleta, Quenstedtoceras lamberti, Quenstedtoceras mariae* and *Cardioceras cordatum*.

The Kellaways Clay is a dark grey, plastic, tenacious clay, up to 10 ft (3 m) thick. The overlying Kellaways Sand consists of light grey coarse silt and sand which may in part be represented by fossiliferous blue-hearted rocks, but which rapidly weathers to a pale yellow colour. The fauna of the Kellaways Beds includes the ammonite *Sigaloceras calloviense* and the bivalves *Pleuromya, Oxytoma* and *Gryphaea*.

The Oxford Clay is a greenish or bluish grey shaly mudstone weathering to a tenacious brown clay. Beds of fine-grained argillaceous limestone, varying in thickness from a few inches to several feet, and calcareous nodules or septaria and small concretionary nodules known as 'race', together with pyrite, occur throughout the formation.

The Oxford Clay contains a rich assortment of fossils, especially ammonites. Commonly associated with these are the annelid *Genicularia vertebralis*, the brachiopod *Aulacothyris bernardina*, the bivalves *Gryphaea dilatata* and *Modiolus bipartitus* and the belemnite *Cylindroteuthis puzosiana* (Pl. IX, fig. 9). The Oxford Clay of the Peterborough area is well known for the remains of numerous genera of large extinct reptiles. They were primarily of marine type and included a variety of plesiosaurs and ichthyosaurs.

After the deposition of the Kellaways Beds, stable conditions throughout Oxford Clay times permitted relatively continuous sedimentation in the eastern part of the district where Brinkmann (1929) has shown the existence of rhythmic deposition in the apparently homogeneous Oxford Clay. At times anaerobic conditions existed on the sea floor. Progressive shallowing brought the bottom into the zone of current action and lamellibranchs and boring organisms were able to live on the sea floor. Subsequently the cycle commenced again with the development of stagnant bottom conditions. Freeman's (1956) chemical and thermogravimetric studies of the Oxford Clay have provided further evidence of its vertical variability.

# 10. Pleistocene and Recent Deposits

The Quaternary Period extends from the end of Pliocene times up to the present day, and may be divided into (1) Pleistocene and (2) Holocene or Recent. The separation is usually made at the end of the Great Ice Age, i.e. at the end of Late Glacial Pollenzone III, but some authorities consider that the Pleistocene extends to the present day.

During this Period a great variety of unconsolidated deposits were laid down, many of which are unstratified or poorly stratified. These beds are grouped under the general term 'Drift' or 'Superficial Deposits' in contrast with the earlier stratified or igneous 'Solid' rocks on which they rest. The Oxford Clay, the youngest surviving 'Solid' formation in Central England, was laid down some 150 million years ago, and in the time-gap between its deposition and the accumulation of the earliest Pleistocene deposits (probably 1 million years ago), many other beds have been laid down, including those of the Cretaceous System, but have been completely removed by erosion.

The beginning of Pleistocene times was marked by a general lowering of temperatures, the formation of snowfields in mountainous regions, and subsequently the growth of ice-caps and glaciers. The ice-masses gradually increased in size and coalesced into huge ice-sheets covering most of north-west Europe. The climate varied continuously but relatively slowly. At times temperate or warm periods intervened, resulting in repeated extensions and retreats of the ice-sheets, so that the Pleistocene can be divided into a series of glacial and interglacial phases.

In the early stages of a glacial phase, the processes of frost-shattering and nivation acting upon the exposed 'Solid' rocks produce an extensive cover of angular material. During periods of seasonal thaw, solifluxion occurs and melt-water streams carry this debris to the lower ground where it is redeposited as fans or floodplain deposits. The advancing ice-sheet incorporates these deposits, together with rock material torn from the floor over which it travels, and transports the whole over considerable distances, boulders and pebbles thus transported being termed 'glacial erratics'. Thus coal and fragments of Carboniferous Limestone from the north of England, Chalk and Upper Jurassic rocks from Yorkshire and the North Sea area, Pre-Cambrian rocks from Charnwood Forest, and foreign rocks, were carried into the South Midlands. During the advance and decay of an ice-sheet this unsorted material is deposited as boulder clay or till. Where this material has been laid down as a thin or irregular blanket of drift it is often called a 'ground-moraine'. If the margin of the ice remains stationary or nearly so for a long period, the rock debris is heaped into a ridge or 'terminal moraine'. 'Kames' are deposits of stratified drift formed in association with ice masses. They have various topographic forms and are thought to have had two major origins; some are thought to have been deposited on or within crevasses in stagnating ice which thawed leaving the isolated gravel mounds: others may have originated as deltas or outwash fans built out from the front of an ice-sheet. Where an ice-sheet impinges on higher ground outwash material from such a standing ice-front is often deposited by water in belts parallel to the ice-front, and forms 'kame terraces', but if the ice-front is retreating or the water has free access to lower ground, streams of melt-water deposit long trails of sand and gravel, known as 'eskers', in channels cut within the ice. Swellings on these eskers mark halt stages where the sub-glacial streams built up delta-fans. During glaciations, melt-water is frequently dammed up, either between separated bodies of ice, or between the ice-sheets and high ground, to form glacial lakes. Outwash material carried into these lakes is thrown down as deltas, the finer-grained material being sorted from the coarse and deposited separately as laminated clays. Plate XB is a photograph of a section in glacial outwash gravels.

**Nomenclature and Classification**

Glacial erratics, and the general constituent material of a boulder clay can indicate the direction from which an ice-sheet came. On the east side of the Warwickshire Coalfield, for instance, Middle Lias debris overlies the Lower Lias outcrop, debris of the latter rests on New Red Sandstone rocks, and New Red Sandstone material is spread over the Coal Measures outcrop, from which it is concluded that the ice in this case travelled from the north-east. Wills (1950) has demonstrated the direction of ice movement in the Midlands by tracing distinctive erratics back to their parent rock outcrops. Chalky till occurs on Charnwood Forest at altitudes of 700 ft (213 m). It would appear that an ice-sheet flowed across these uplands from the north-north-east carrying the Charnian erratics in a southerly direction. Thus the ice must have been at least 750 ft (230 m) thick in Leicestershire, and at its maximum extent in the Midlands it may have been considerably thicker.

During glacial phases Central England was liable to invasion by ice from four directions: from the Welsh Mountains on the west, from the Irish Sea Basin on the north-west, the Pennines in the north and from the Vale of York and the North Sea Basin on the north-east. The drift deposited by an ice-sheet has a characteristic suite of erratics and this is used as a basis of classification. Thus the drift deposits laid down by ice-sheets from Wales and from the North Sea Basin are defined as Western and Eastern Drift respectively, whilst those deposits derived from Scotland and the Lake District via the Irish Sea Basin are referred to as Irish Sea Drift. In South Warwickshire, at the southern extreme of the Central England Region, drift characterized by an abundance of pebbles, probably derived from the Bunter Pebble Beds of the Midlands, has been described as Northern Drift. The erratic content of a boulder clay has no direct age significance, but during a particular period, streams or layers of ice containing erratics from a specific source may have been dominant locally. Thus division on the basis of erratic provenance is unsuitable. The order of superposition of glacial deposits can be used locally, but it cannot be extended with certainty. The ideal basis would be the creation of time-stratigraphic units but because of the variable nature of the deposits and the general lack

TABLE 2. Regional succession of the Quaternary deposits of Central England and East Anglia

| | | | Stage | Climate | Central England Region | East Anglia |
|---|---|---|---|---|---|---|
| Quaternary | Holocene | | Flandrian | t | Peat, alluvium, raised beaches, estuarine clays | |
| | Pleistocene | Upper | Weichselian | c | Worcester Terrace (Severn) Avon 1 Terrace Main Terrace (Severn) Avon 2 Terrace Main Irish Sea Glaciation Chelford organic deposit | Solifluxion deposits and patterned ground<br><br>Hunstanton Till |
| | | | Ipswichian | t | Kidderminster Terrace (Severn) Avon 3 and 4 Terraces | Interglacial deposits of Cambridge, etc. |
| | | | Gipping | c | Bushley Green Terrace (Severn) Avon 5 Terrace Main Eastern Glaciation (Chalky Boulder Clay & Dunsmore gravels, Wolston Series, Baginton-Lillington gravels) 2nd Welsh Glaciation Older Irish Sea Glaciation | Gipping Till |
| | | Middle | Hoxnian | t | Interglacial deposits at Nechells (Birmingham) | Interglacial deposits at Clacton, Hoxne, etc. |
| | | | Lowestoft | c | 'Bubbenhall clay' 1st Welsh Glaciation Northern Drift | Lowestoft Till Corton Beds North Sea Drift |
| | | Lower | Cromerian | t | Deposits restricted to East Anglia | |
| | | | Baventian | c | | |
| | | | Antian | t | | |
| | | | Thurnian | c | | |
| | | | Ludhamian | t | | |

c = cold; t = temperate

(Published by permission of The Geologists' Association)

of fossils within them this can rarely be achieved. West (1963) has suggested that time-stratigraphic units can be established on the basis of climatic variations which are reflected in land forms, the nature and lithology of sediments and the fauna and flora.

**Sequence**

The sequence of advance and retreat of ice-sheets may have been more or less contemporaneous throughout Europe, but the duration and extent of the movement depended upon local conditions of climate and relief. Although four periods of maximum glaciation are thought to have been recognized in Continental Europe, the interpretation of the Pleistocene succession in Britain is very far from complete. Most geologists consider that there is evidence of three glacial episodes in this region. The last episode, being divisible into stages of re-advance, has sometimes been divided into a further two glaciations. West (1963) has summarized these views as shown in Table 2.

Wills (1948) grouped the deposits of the first two glaciations (Lowestoft and Gipping) together as the Older Drifts. These deposits have a patchy distribution in the Midlands and are usually confined to the higher ground. The glacial tills and sands and gravels of these episodes are separated by a well-marked period of erosion from the deposits of Newer Drift age. The latter are river terraces in the Midlands and glacial sands and gravels and tills in the Shropshire–Cheshire Basin. Although deposits attributed to the Older Drift glaciations, i.e. the First and Second Welsh Glaciations and the Older Irish Sea Glaciations, have been recognized in the Midlands, none can be certainly recognized in the Shropshire–Cheshire Basin.

However, Poole and Whiteman (1961) considered that the Shropshire–Cheshire sequence is best explained as resulting from two major glaciations corresponding to an Upper and a Lower Boulder Clay respectively. During each of these periods boulder clay, glacial sands and gravels, terminal moraines, glacial lake features and melt-water channels were produced. The limit of the Main Irish Sea Glaciation as defined by Wills (1948) is regarded by them as the boundary of a till deposit of distinctive type with a highly characteristic erratic suite which at the maximum of the Upper Boulder Clay Glaciation marked the line where the Irish Sea ice-sheet impinged against ice-sheets from Wales and the Pennines. This implies that the deposits of the Upper Boulder Clay Glaciation correlate directly with those of the Second Welsh Glaciation and consequently that the Lower Boulder Clay Glaciation was contemporaneous with the First Welsh Glaciation. The apparent anomaly of correlating Older and Newer Drift deposits is explained by the diachronous nature of all glacial deposits. Their interpretation of the Pleistocene sequence of the Shropshire–Cheshire Basin and adjacent areas is shown in Table 3.

Current work in the Macclesfield district (Evans *et al.* 1968) in the eastern part of the Shropshire–Cheshire Basin has confirmed the existence of at least two glacial periods. The general drift sequence is similar to that established by Poole and Whiteman though no regional Lower Boulder Clay is recognized, but an alternative hypothesis for the origin and age of the deposits is proposed. The only evidence of the early glacial advances are

## Pleistocene and Recent Deposits

TABLE 3. GENERALIZED PLEISTOCENE SEQUENCE AND DEPOSITS OF THE SHROPSHIRE–CHESHIRE BASIN
(After Poole and Whiteman 1961)

[Table content - a large stratigraphic chart oriented sideways, showing from North to South:]

**GEOLOGICAL SURVEY ONE-INCH SHEETS:**
83 (Formby); 84 (Wigan); 85 (Manchester); 108 (Flint); 109 (Chester); 110 (Macclesfield); 137 (Oswestry); 138 (Wem); 139 (Stafford)
96 (Liverpool); 97 (Runcorn); 98 (Stockport); 121 (Wrexham); 122 (Nantwich); 123 (Stoke); 152 (Shrewsbury); 153 (Wolverhampton)

**TIME STRATIGRAPHIC UNITS** and **RADIOCARBON DATING:**

| Deposits / Events | Time Stratigraphic Units | Radiocarbon Dating |
|---|---|---|
| ALLUVIUM, PEAT, SHELL-MARL ETC. THROUGHOUT BASIN. FIRST AND SECOND TERRACES OF RIVERS DEE, MERSEY AND WEAVER | RECENT OR POST GLACIAL PHASE | 0 |
| TILLS AND OUTWASH DEPOSITS OF SCOTTISH READVANCE PHASE NOT REPRESENTED IN THE BASIN BUT LATE GLACIAL DEPOSITS PROBABLY OCCUR IN CERTAIN PEAT BOGS | LATE GLACIAL AND SCOTTISH READVANCE | 10,000 Years |
| MAXIMUM EXTENSION OF RETREAT WELL NORTH OF SHROPSHIRE–CHESHIRE BASIN | MAXIMUM RETREAT PHASE | |
| ICE-FRONT SITUATED ACROSS DEE ESTUARY WITH MINOR FLUCTUATIONS. FLUVIOGLACIAL TERRACES AND DEPOSITS OF RIVERS LEE, MERSEY, WEAVER, PENK, ETC. | DEE ESTUARY HALT PHASE | |
| ICE-FRONT EXTENDING N–S ALONG MID-CHESHIRE AND UPHOLLAND RIDGES WITH MINOR FLUCTUATIONS. LAKE LAPWORTH STAGES. 130 FT, 170 FT, 200–220 FT — FIRST TERRACE OF RIVER SEVERN / FIRST (ATCHAM) TERRACE OF RIVER SEVERN; 250–260 FT — SECOND (CRESSAGE) SEVERN TERRACE / THIRD (UFFINGTON) SEVERN TERRACE / MAIN SEVERN TERRACE AND TRENT FLUVIOGLACIAL GRAVELS (IN PART); 305 FT, 330 FT | MID-CHESHIRE RIDGE HALT PHASE | |
| ICE-FRONT ALONG WESTERN SIDE OF PENNINES WITH MINOR FLUCTUATIONS. LAKES OF CANNOCK CHASE, MADELEY, OAKENGATES AND CALBROOKDALE; TRENT FLUVIO-GLACIAL GRAVELS (IN PART) | PENNINE HALT PHASE | |
| BASIN FILLED WITH ICE; IRISH SEA ICE-SHEET CONFLUENT WITH WELSH, PENNINE AND EASTERN ICE. PROBABLY GREATER THAN 1,200 FT THICK. MAXIMUM EXTENSION WELL BEYOND BASIN SOUTH-EASTWARDS | UPPER BOULDER CLAY ADVANCE AND MAXIMUM EXTENSION PHASES | UPPER SANDS |
| MAXIMUM EXTENSION OF RETREAT WELL NORTH OF SHROPSHIRE–CHESHIRE BASIN | MAXIMUM RETREAT PHASE | 40,000 Years |
| ICE-FRONT SITUATED AT THE CHORLEY MORAINE. ICE-STANDS OF DELAMERE AREA | CHORLEY HALT PHASE | |
| ICE-FRONT SITUATED AT THE WREXHAM-BAR HILL MORAINIC SUITE WITH FORMATION OF WREXHAM DELTA TERRACE. FORMATION OF −300 FT MIDDLE SANDS LAKE. PROBABLE DATE OF INITIATION OF IRONBRIDGE GORGE AND MADELEY GAP | WREXHAM–BAR HILL HALT PHASE | |
| NEWPORT–WOLVERHAMPTON ESKER CHAIN | EARLIER UNDIFFERENTIATED PHASES | MIDDLE SANDS |
| BASIN FILLED WITH ICE; IRISH SEA ICE-SHEET CONFLUENT WITH WELSH, PENNINE AND EASTERN ICE. PROBABLY GREATER THAN 1,200 FT THICK. MAXIMUM EXTENSION WELL BEYOND BASIN SOUTH-EASTWARDS | LOWER BOULDER CLAY ADVANCE AND MAXIMUM EXTENSION PHASES | 57,000 Years |

UPPER BOULDER CLAY / LOWER BOULDER CLAY

ERODED SURFACE OF PRE-PLEISTOCENE ROCKS

remanié patches of boulder clay but deposits of the last glaciation are much more varied and extensive.

Is it coincidence that only two glacial sequences have been recognized in each of the areas of detailed study within Central England? In south Warwickshire, Shotton (1953) has recorded remnants of a deposit with till-like affinities. This, the Bubbenhall clay, is said to be overlain by deposits which include Chalky Boulder Clay. A sequence showing Chalky Boulder Clay overlying glacial gravels and an older till has been recognized in Northamptonshire (Hollingworth and Taylor 1946). Throughout most of East Anglia only two till sheets are present (West and Donner 1956), but near Hunstanton a third and younger till is said to be developed. One critical factor in the controversy is the age of the Chelford organic deposit (Simpson and West 1958) which is said to be 57,000 years B.P. (Weichselian). Poole and Whiteman do not accept this date, as they consider that radiocarbon dating cannot be extended accurately much beyond 40,000 years B.P. However radiocarbon dating of material from within the deposits has confirmed the correlation of the Main Terrace of the Severn (42,000 years B.P.) and the Second Terrace of the River Avon (39,000 years B.P.). Since the publication of Table 3 radioactive carbon dating of material in gravels underlying the Upper Boulder Clay of Cheshire has shown that this till is younger than the Main Terrace of the Severn but the problem of correlation of the Older and Newer Drifts is still unsolved.

A further difficulty is raised by the identity of the 'Bubbenhall clay' at the type locality of Bubbenhall, near Coventry. Evidence from recent shallow borings suggests that the succession in the basal glacial deposits of this area needs reconsideration. One important area for future study lies between Birmingham and the Ironbridge Gorge. Because these problems are as yet unsolved the Shropshire–Cheshire Basin and the Midlands are considered separately.

*Central England Region (excluding the Shropshire–Cheshire Basin)*

Scattered deposits of pre-glacial sands and gravels, which are younger than the local glacial deposits, have been recorded from Northamptonshire (Hollingworth and Taylor 1946).

At the commencement of the first glaciation various ice-sheets advanced from the north and west depositing till over these gravels of local origin. The lithological characters of the tills vary and they have been described as Western Drift or Eastern Drift and have been correlated with the Northern Drift or Plateau Gravels of the Cotswold–Oxford area. Wills considered that this glaciation (the First Great Welsh Glaciation, Lowestoft, Mindel or Elster) stopped at the Jurassic escarpment; but it is generally accepted that it overrode the escarpment, and Arkell suggested that the southern limit may have been against the hills of the Vale of White Horse.

The climate ameliorated, the ice decayed in Central England and a great period (Hoxnian) of erosion and fluviatile deposition occurred. Few of these sediments are preserved but the Nechells peaty silts of the Upper Tame Valley are said to belong to this interglacial stage (Duigan 1956).

Subsequently, ice-sheets re-entered the area and the Second Great Welsh (Great Eastern, Gipping, Saale or Riss) Glaciation commenced. The Welsh ice is said to have been much weaker and therefore much less extensive than in the previous glaciation; consequently the bulk of the drift in the west was derived from Scotland and the Lake District and was carried into the Midlands via the Cheshire Plain by the Irish Sea ice. A powerful ice-sheet from the north-east, the Great Eastern, brought an abundance of flint and chalk, Jurassic debris and rocks from Charnwood and the Pennines and deposited the Chalky Boulder Clay. The two great ice-sheets impinged along a line running approximately from Derby to Coventry and thence to Stratford. The Great Eastern ice-sheet was more powerful and reached as far as Tewkesbury and overrode the Moreton Gap. For the second time the whole region was covered by ice.

When the ice retreated, Central England had a long period (Ipswichian) of temperate or warm climate. *Hippopotamus* and other animals which today live in warmer climes, roamed the flood plains of the aggrading rivers. The Kidderminster Terrace of the Severn, the third and fourth terraces of the Avon and the Hilton and Beeston 'terraces' of the Trent belong to this period.

Wills has correlated the Main Terrace of the Severn with the third or Main Irish Sea Glaciation (Weichselian) and the Worcester Terrace of the same river with a fourth glacial episode, the Welsh Re-advance. The terraces of the other rivers have been correlated with these glacial periods.

The Quaternary succession of the Coventry–Rugby–Leamington area has been described by Shotton (1953) as follows:

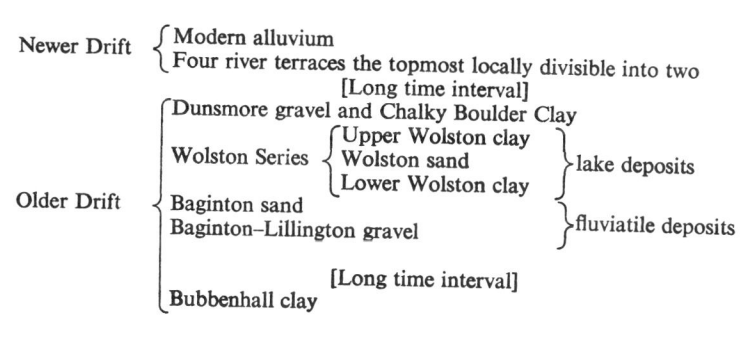

```
Newer Drift  { Modern alluvium
             { Four river terraces the topmost locally divisible into two
                              [Long time interval]
             ┌ Dunsmore gravel and Chalky Boulder Clay
             │                  ┌ Upper Wolston clay  ┐
             │  Wolston Series  { Wolston sand         } lake deposits
Older Drift  {                  └ Lower Wolston clay  ┘
             │  Baginton sand                          ┐
             │  Baginton–Lillington gravel             } fluviatile deposits
             │                                         ┘
             │           [Long time interval]
             └ Bubbenhall clay
```

He suggested that the Baginton–Lillington gravels were laid down in a strike valley (the proto-Avon valley) on the Keuper Marl, which sloped gently down to the north-north-east from Bredon Hill, across the present watershed of the Avon, to Leicester, and formed part of the Trent system. The fauna of the gravels indicates a cold climate which suggests that deposition was occurring at the commencement of the second glaciation. As the Great Eastern ice-sheet advanced from the north the outlet was closed. The gravels merge up into the Baginton sand which contains foreset beds and at the top, interbedded silts, suggesting that ponding must have commenced during their deposition. As the ice-sheets advanced they impounded the rivers to form Lake Harrison which extended over a wide area of the Midlands, from a few miles north of Leicester to Market Bosworth, across to Birmingham and south to Moreton-in-Marsh, a distance of 56 miles (90 km). The Wolston Series was laid down in this lake. Pebbles occur throughout the clays: in the lowest deposits these consist of Bunter and Keuper Sandstone erratics, but later beds contain up to 50 per cent of flint. Shotton suggested that a northern ('western') ice-sheet entered the area first and formed the lake, but later it was pushed aside by ice advancing from the north-east. The latter, the Great Eastern ice-sheet, advanced across the earlier lake deposits and reached as far as Moreton-in-Marsh. During the melting of the ice-sheet part of the earlier-formed deposits was eroded away. In the Dunsmore area its melt-waters laid down the Dunsmore gravels.

Bishop (1958) has confirmed that the history of Lake Harrison was complex. He has recognized two major stages of ponding which were separated by the maximum extension of the Chalky Boulder Clay ice-sheet. The advance of the ice-sheet was progressive but continuously oscillating. At first the water level must have been at least 435 ft (132·6 m) but at the maximum stage the level must have fallen to 415 ft (126·5 m) A.O.D. When the ice retreated from the Moreton col the lake may have been as high as 430 ft (131·1 m) and the melt-water spilling into the Evenlode is said to have deposited gravels which have been correlated with the Wolvercote Terrace of the Thames. At a later retreat stage, the surface was at about 410 ft (125 m) A.O.D. and water escaped into the Cherwell via the Fenny Compton spillway. Dury (1951) reported a discontinuous 35 mile (56·3 km) lake bench in this part of south-east Warwickshire, which was produced at this time.

Shotton suggested that ice coming up the Tame Valley would impound lakes between itself and the Clent Hills. This has since been confirmed by Pickering (1957) who recognized lake deposits in the Birmingham area which attain a maximum height of 545 ft (166 m) A.O.D. At the end of the Second Glaciation (Wills 1948), deposits of Chalky Boulder Clay filled the pre-glacial proto-Avon valley between Nuneaton and Rugby, causing the rivers to flow south-westward into the Severn.

*The Shropshire–Cheshire Basin and the country north-west of Wolverhampton*

Wills (1951) considered that the 'Newer Drifts' of this area were laid down during the last glaciation. Three cold stages separated by warmer periods (interstadials) are said to have been recognized. The latter have been claimed to be true interglacial periods, but are now considered to be only stages of retreat during a single glaciation. The sequence is said to be: (i) maximum of Little Eastern and Main Irish Sea Glaciations; (ii) retreat with the formation of marginal lakes Coalbrookdale, Newport etc.; (iii) Little Welsh Glaciation and the Ellesmere Moraine-belt stage of the Main Irish Sea Glaciation; (iv) retreat stage and the (v) Scottish Re-advance.

Poole and Whiteman (1961, 1966) described the sequence in the Nantwich district and applied it to the whole of the Shropshire–Cheshire Basin. Their sequence in ascending order is: (i) Lower Boulder Clay Glaciation; (ii) Middle Sands outwash and lacustrine deposits formed at successive retreat stages of the Lower Boulder Clay ice-sheets; (iii) Upper Boulder Clay Glaciation; (iv) fluvioglacial and lacustrine deposits (marginal lakes of Cannock Chase, Coalbrookdale, Oakengates, Newport and Lake Lapworth) and (v) terraces, peat and alluvium. These deposits are said to be the result of two periods of glaciation during which the two boulder clays were formed. During each of the subsequent retreat phases, pre-glacial lakes were impounded, and they demonstrated how the present-day drainage of the basin evolved from the melt-water features produced during the retreat of the Upper Boulder Clay ice-sheet.

Both theories are based on the existence of evidence for at least two glaciations in the area, but they differ in the correlation of this sequence with that in the Midlands. Wills considered that the area was subjected to three glaciations, the deposits of the earlier two, the Older Drifts, having been completely removed or incorporated in the boulder clay of the last glaciation. Poole and Whiteman considered that the drifts of the Shropshire–Cheshire basin (Wills' Newer Drifts) were formed during two glaciations and are lateral equivalents of the Older Drifts to the south-east.

Wills limited the Irish Sea Glaciation in the vicinity of Wolverhampton, whereas the others agreed that although this was its maximum extent, it here impinged upon ice-sheets from Wales and the Pennines (Eastwood and others 1925, p. 106). Dixon (*in* Whitehead and others 1928) has recorded successive ice-stands and the formation of the Newport–Wolverhampton esker chain which were produced during the retreat of the Irish Sea ice-sheet. The esker is said to have been laid down by a melt-water stream issuing from the base of the ice-sheet. When the ice had retreated behind the Wenlock–Pennine watershed, the gravels were deposited as deltas in the lake formed between the ice-sheet and the watershed (Lake Newport). Recent research has shown that gravels belonging to the 'esker' are in fact overlain by till, and hence Poole and Whiteman argue that they could not have been formed by the final retreat of the Irish Sea ice as envisaged by Dixon and Wills. The sequence is seen as confirming the evidence of two glaciations in the area north-west of Wolverhampton. They conclude that the 'esker' gravels were deposited during the retreat of the earlier ice-sheet and though they may have in part been deposited within a lake it was not Lake Newport, which was formed by the final retreat of ice from the area. However, superposition of till on gravel could have arisen if the esker was sub- or englacial and the till was deposited when the overlying or enclosing mass of dirty ice melted.

Poole and Whiteman believe that as the earlier ice-sheet retreated farther the Middle Sands were laid down throughout the Shropshire–Cheshire Basin. They consist mainly of sand with minor deposits of gravel, silt and laminated clays which were thought to have been deposited as moraines, eskers, outwash fans and lacustrine clays.

During the north-westward retreat of the last Irish Sea ice (Upper Boulder Clay Glaciation of Poole and Whiteman, 3rd Glaciation of Wills) from the Shropshire–Cheshire Basin, melt-water including that of the present Upper Severn drainage was impounded between the ice-front and the Wenlock–Pennine watershed to the south and east. Wills suggested that one of the earliest lakes, Lake Coalbrookdale, drained across Lightmoor along the valley of the Coalport Brook; later this channel was blocked and the water, diverted across the Ironbridge col, commenced the cutting of the Ironbridge Gorge. Erosion lowered the outlet and retreat of the ice resulted in an increase in the area of water to form Lake Buildwas. At this time, Lake Newport occupied an embayment in the hills to the north-east. With further retreat of the ice-sheet the two bodies of water coalesced to form Lake Lapworth.

Poole and Whiteman considered that these lakes were much more extensive than originally thought. They suggested that after the maximum of the Upper Boulder Clay Glaciation, the ice retreated from the Midlands. Colder conditions prevailed when the retreating ice-front lay along a line joining the Pennines and the Ironbridge Gorge. Marginal lakes were formed in several areas including Madeley (Yates and Moseley 1957), Cannock and Ironbridge (Lakes Coalbrookdale and Buildwas), and at this time the Kidderminster Terrace was probably formed. With a renewal of warmer conditions the ice-front withdrew to a position in the centre of the Shropshire–Cheshire Basin and impounded a lake at 330 ft (100·6 m) which they believe extended from Manchester to Shrewsbury. This is the level of Wills' Lake Newport. Probable overflow points were (i) west of Eccleshall and (ii) at Gnossall both into the Trent basin and (iii) at Ironbridge into the Severn. Rapid erosion at the southern outlet at Ironbridge lowered the lake level to about 305 ft (93 m) (Lake Lapworth) and all the water drained into the Severn. They correlate the Main Terrace of this river with these two lake stages. Downcutting proceeded rapidly in the Ironbridge Gorge, possibly due to the re-excavation and deepening of an older valley which was originally filled with unconsolidated deposits similar to those now remaining at Buildwas. In the later stages of the formation of the Ironbridge Gorge the whole area south of the Wrexham–Whitchurch Middle Sands morainic ridge was drained. Further retreat or lowering of the ice barriers led to the impounding of lakes at 260 to 250 ft (79·2–76·2 m), 220 to 200 ft (67–61 m), 170 to 150 ft (51·8–45·7 m) and at 130 ft (39·6 m), behind this ridge. The First Terrace of the Severn north of Ironbridge is correlated along the valley of the River Tern to the melt-water gap at Adderley which corresponds with the 220 to 200 ft (67–71 m) lake stage. The Second and Third Terraces in this area probably correlate with the 250 ft (76·2 m) and 260 ft (79·2 m) lake stages. After minor oscillations the ice-sheet is thought to have retreated to a line across the estuaries of the rivers Dee and Mersey and impounded a lake at 130 ft (39·6 m) in the northern parts of the Shropshire–Cheshire Basin. The melt-waters escaped along the edge of the Welsh Mountains and drainage to the Irish Sea was re-established. With the final retreat of the ice from the Basin the last cycle of downcutting began and the terraces and alluvium of the rivers Dee, Mersey and Weaver were formed.

In the Macclesfield district (Evans *et al.* 1968) remanié patches of boulder clay lie at the base of the local Quaternary sequence. They are overlain by a deposit of sand referred to as the Congleton Sand and the Chelford Sand, which is succeeded by the Gawsworth Sand, and the Upper Boulder Clay. A suite of fluvioglacial sands and gravels and a spread of laminated clays, possibly of glacio-marine origin, complete the Pleistocene sequence.

The Congleton–Chelford Sand and the Gawsworth Sand are divisions of the Middle Sands of earlier authors. The lower unit, the Congleton Sand and its equivalent the Chelford Sand, consists largely of water-lain wind-blown sand derived from the Pennine area. Its flora and fauna at Chelford suggest that the region was ice-free during its formation. The upper unit, the

Gawsworth Sand, is best developed between Macclesfield and Nantwich along the margin of the Shropshire–Cheshire Basin. It is thought to be a morainic deposit partly of sub-glacial origin associated with the ice-sheet which deposited the Upper Boulder Clay during the last glaciation of the district. This interpretation more readily explains the relatively unmodified topography of these morainic deposits than the Poole and Whiteman hypothesis which considered that their Middle Sands was morainic outwash formed during the retreat of the penultimate Lower Boulder Clay ice-sheet and which was preserved due to its being frozen during the advance of the subsequent ice-sheet.

Accepting the radiocarbon dating of the organic horizon in the Chelford Sand, it is argued (Evans *et al*. 1968) that the younger till must be of Main Weichselian age. This has been confirmed by a carbonate dating of about 28,000 years B.P. for two molluscan shells found in gravels underlying the Upper Boulder Clay in the Sandiway district of Cheshire (Boulton and Worsley 1965). Peaty deposits in gravels underlying the Upper Boulder Clay at Four Ashes, five miles north of Wolverhampton, have been dated at about 30,000 and 36,000 years B.P. respectively. Thus during Main Weichselian (late Würm) times the Irish Sea Glacier extended at least to the vicinity of Wolverhampton (Shotton 1967). The final stages of Lake Lapworth have long been correlated with the Main Terrace of the Severn and its correlative the No. 2 Terrace of the Avon which together have yielded material dating from 28,000 to 42,000 years B.P., i.e. to the time of the Upton Warren (Mid Würm) Interstadial. Thus there is a conflict between the suggestion that the highest Lake Lapworth features are cut into the Upper Boulder Clay and the downstream correlation of these features with the Main Terrace.

Officers of the Geological Survey have found no evidence of subsequent penetration of ice-sheets into the Chester area at the time of the Scottish Re-advance. After the decay of the Upper Boulder Clay ice-sheet 'Glaciers did not enter the Shropshire–Cheshire Basin again, the ice of the Scottish Re-advance glaciation being restricted to the northern end of the Isle of Man . . . and the Lake District' (Poole and Whiteman 1961, p. 120).

**Periglacial Deposits**

Periglacial deposits are formed in the arctic or sub-arctic climatic zone surrounding an ice-sheet. Intermittent expansion and decay of the ice sheet in the Midlands resulted in a diachronous relationship between the intimately associated glacial and periglacial deposits, while in warmer periods interglacial deposits accumulated. The classification of deposits formed during each of these stages is extremely difficult.

*Fluvioglacial sand and gravel*. Melt-water streams associated with ice-sheets are loaded with debris, which is rapidly deposited as coarse ill-bedded spreads of fluvioglacial sand and gravel. Much of this material is laid down in front of the ice where it is liable to be removed by the melt-water, but it is also deposited within or beneath the ice or even on its upper surface. Deposits of this type, mainly irregular masses of sand and gravel, are found intimately associated with the tills of Central England.

More distant from the source, fluvioglacial gravels merge imperceptibly into river terrace deposits. Extensive deposits of gravel occur on the north sides of the rivers Dove and Trent. These show a great variation in height above the present alluvium and were described as fluvioglacial gravels (Stevenson and Mitchell 1955). Posnansky (1960) concluded 'that they are outwash aggradation terraces of the melting ice of the Eastern Glaciation'. The two lowest levels of this series form two well defined terrace 'flats'.

The Main Terrace of the Severn is contemporaneous with the cutting of the Ironbridge Gorge (or alternatively its re-excavation and deepening during the final retreat of the Irish Sea ice-sheet) and is therefore periglacial. The lower terraces of the Severn north of the Gorge and the Worcester Terrace to the south have also been correlated with stages in this retreat.

*Head.* Alternate freezing and thawing causes frost-shattering of rocks, the debris forming scree or talus slopes. When mixed with ice or clay and silt grade material the whole may form 'stone rivers' or mud flows. Unsorted debris of this kind, usually described as Head, surrounds the crags of slaty rock in Charnwood and mantles the slopes beneath the basalt which caps Titterstone Clee Hill. To the west and south-west of the Clent Hills, extensive spreads of breccia gravels are present. They consist of rock fragments set in a sandy matrix and were probably initiated as solifluxion deposits, i.e. weathered material saturated with water and moving downhill under gravity. These flows may have been re-sorted by water later, since they grade into river terraces.

In the south-west of the Region, 'taele' gravels have been deposited on the low ground below the Cotswold escarpment. They consist largely of local limestone debris. The highest deposits are unstratified and show evidence of solifluxion, but downstream they grade into terraces correlated with the Main Terrace of the Severn.

In the Northampton Sand Ironstone Field, Head, consisting of sandstone, ironstone and limestone rubble, frequently masks many of the slopes and forms stony and silty deposits flooring small dry valleys. Many of the east Midland river terraces pass laterally into deposits of angular or subangular material which cannot readily be distinguished from Head.

*Wind-etched stones.* During the last cold phase of the Pleistocene period, conditions in the Midlands were similar to those of the cold and dry steppelands bordering glaciated regions at the present time. Strong winds blowing across the area caught up the finer parts of the glacial drifts and gave rise to sandstorms. Blocks of stone lying on the surface were etched and polished by sandblast action to form wind-facetted pebbles, many of which have been found in the superficial deposits at Lilleshall and around Droitwich. The wind-polished surfaces seen on the granitic rocks of Mountsorrel are also thought to be of Pleistocene age (Raw 1934).

*Patterned ground.* Shotton (1960) has described polygonal markings on the surface of No. 4 Avon Terrace, near Evesham, and considered that they were formed under permafrost conditions. Patterned ground is also known from Northamptonshire and the Fens.

## Post-Glacial and Recent Deposits

*River terraces and alluvium.* These are the most important and extensive of the Post-Glacial deposits. The difficulty of separating these deposits from those associated with the last glaciation has been mentioned earlier. In the Severn basin two or possibly three gravel terraces later than the Main Terrace have been identified. The upper two have been correlated with the Worcester or Second Terrace of Wills (1937). Five terraces have been recognized in the Avon valley; the second terrace is correlated by both Shotton (1953) and Bishop (1958) with the Main Terrace of the Severn and hence with a stage in the final retreat of the Irish Sea ice-sheet; and the fifth terrace with the closing stages of the Second Welsh or Chalky Boulder Clay glaciation. The relative age of these terraces is uncertain; Tomlinson (1925) and Shotton (1953) suggested a sequence 5, 3, 4, 2, 1, but in the Banbury district (Edmonds *et al.* 1965) the simple age sequence 5, 4, 3, 2, 1 with the highest terrace (No. 5) being the oldest is favoured. In the Trent valley two terraces younger than the Hilton terraces have been recognized. In the east Midland rivers, notably the Nene and the Welland, three main terraces are recognized. These post-date the Chalky Boulder Clay but their correlation with the Severn sequence is as yet uncertain.

Alluvium is present along the banks of most rivers where it forms low-lying areas liable to flooding. It consists of silty clay with occasional thin gravel bands and layers of pebbles associated with freshwater shells and the bones of modern animals. Near the sources the alluvium contains a higher proportion of gravelly material.

*Fen deposits.* The low-lying ground north of Peterborough forms a small part of the Fenlands. The deposits, which locally overlie boulder clay, attain a known thickness of 60 ft (18 km) and comprise silts and clays with beds of peat. In some areas beds with marine fossils were laid down. The rivers, including the Welland and the Nene, brought loads of fine-grained material which was deposited in an estuarine environment. In the regions most distant from the sea peat formation occurred almost continuously, but seaward, fluctuations in relative sea level interrupted the growth of plants and today the beds of peat are seen to interdigitate with estuarine silts and clays. Sand and silt were deposited in the beds and as levées along the sides of streams. Differential compaction of the peat formed in the interdistributary areas, aided by a lowering in water level, has resulted in the coarser deposits standing out as ridges above the surrounding countryside. Deposits of *Chara* Marl laid down in relict meres are amongst the most recent fen deposits.

*Peat.* In South Cheshire and Shropshire meres occupy hollows in the irregular sheet of Drift left by the ice of the last advance from the Irish Sea. Some of these have silted up and been converted into peat bogs, for example on Whixall Moor and near Ellesmere, where peat cutting was an important local industry. Radiocarbon dating of organic material underlying a peat moor at Rodbaston, near Penkridge has given an age of $10{,}670 \pm 130$ years (Shotton and Strachan 1959).

*Tufa and shell marl.* Small spreads of calcareous tufa formed by the deposition of calcium carbonate from springs have also been recorded from several places in this region. Beds of shell marl have accumulated in certain of the relict meres of Cheshire.

## Changes in River Drainage

The drainage of Central England is said to have originated by superposition from a cover of Mesozoic rocks which tilted to the south-east. A series of consequent streams were developed which flowed in this direction. When erosion revealed the Jurassic outcrops, tributary streams are thought to have cut into the clay outcrops to produce broad vales, and in time some of the consequent streams were beheaded. Wills (1950) considered that gentle faulting and folding of the Mesozoic cover may have modified this consequent drainage (if it existed); uplift of the present coalfield areas produced extensive downwarps along which the earliest rivers flowed. He recognizes three major basins, (i) the Upper Severn flowing into the Cheshire basin, (ii) the Avon and its tributary the Middle Severn, and (iii) the Trent. Linton (1951) suggested that the pre-Eocene earth-movements produced a land surface on the Chalk which tilted to the east. Two major consequent streams developed, the proto-Thames (along the Kennet–Thames line) and the proto-Trent (Dee–Trent), the two systems being separated by a watershed running from Plynlimon to Birmingham and Market Harborough. The folding of the Lias in the Cheshire Basin appears to contradict the eastward tilt postulated by Linton but work in the basins of the Welland and Nene (Kellaway and Taylor 1953) has shown that in this area the present erosional cycle commenced on a surface gently sloping towards the North Sea.

The present drainage system includes some partly re-excavated pre-Glacial valleys and others in which there are small deposits of what have been called 'pre-glacial' gravel. Many of the older valleys were blocked with till or glacial gravel, and melt-water streams cut new channels. Shotton (1953) has shown that at the start of the Riss or Main Eastern glaciation the Middle Severn had at most a very rudimentary tributary along the valley of the Avon. Much of Warwickshire and Worcestershire was at that time drained by a river which flowed north-eastward through a large valley into the Trent system. To the south-east the Jurassic escarpment formed the watershed between this valley and the Thames. Melt-waters from the retreating ice-sheets enabled the primitive Avon to cut back rapidly through the soft Keuper Marl and drift deposits, and the present drainage pattern was established before the aggradation of the oldest terrace began.

Wills (1912) has shown that the upper reaches of the Severn originally drained into the Irish Sea via the River Dee. In Pleistocene times, this route was blocked by the advancing ice-sheets during two or more periods of glaciation. Within each of these periods the impounded water escaped to the south and east; during the last or possibly the last two glaciations the Ironbridge Gorge was cut. When the ice-sheet retreated the northern drainage was blocked by till deposits, hence the waters of the Upper Severn which had flowed through the Cheshire Plain were diverted into the Middle Severn and thence to the Bristol Channel. Posnansky (1961) suggested that the pre-glacial Trent flowed into The Wash, but during the retreat of the Eastern Glaciation ice-sheets it was diverted northward and escaped first through the Lincoln Gap before establishing its present course into the Humber. Other buried gaps of comparable size exist in the drift-covered areas to the south, e.g. between South Witham and Castle Bytham in Lincolnshire.

Drift-filled valleys are known from borings and opencast workings in the Northampton Sand Ironstone Field and the Midland coalfields. Many of the existing rivers flow along drift-filled channels for part of their course.

# 11. Economic Geology

In comparison with other parts of Great Britain, a large proportion of the land within Central England is devoted to urban development. Nevertheless, farming is an important industry, usually with mixed arable and animal husbandry predominating. In Cheshire and Shropshire, dairy farming predominates, whereas mixed farming characterizes the outcrop of the Jurassic rocks. The Vale of Evesham is famous for its fruit growing. Market gardening is important in the country south of Lichfield and in the Fenlands, which are the most intensively cultivated part of Central England.

Coal, limestone, ironstone, brick and pottery clays and sand and gravel are the major economic products. Others which are or have been worked are oil, rock salt, gypsum, building stones, road-stones and aggregates, refractories and ganisters, moulding sand and ores of copper, manganese and lead.

### Coal

In 1965/66 saleable output of coal from the seven coalfields in the region exceeded 20 million tons. The amounts raised in the individual coalfields were: North Staffordshire 5,379,182 tons, South Staffordshire (including Cannock Chase) 3,683,308 tons, Coalbrookdale 302,745 tons, Wyre Forest 240,840 tons, Warwickshire 3,822,378 tons, South Derbyshire 2,683,983 tons and Leicestershire 4,226,913 tons.

### Limestone and Cement

The bulk of the limestone quarried in this region is obtained from the Jurassic limestones of Warwickshire, Northamptonshire and Rutland. The manufacture of cement is an important industry; the interbedded limestones and clays of the Blue Lias provide suitable raw material, and they are worked at Southam, near Leamington, at Rugby, and at Barnstone, east-south-east of Nottingham. At Ketton, south-west of Stamford, the Lincolnshire Limestone is mixed with the Upper Estuarine Series and used in the manuacture of cement. The Lincolnshire Limestone has been dug for the same purpose at Waltham on the Wolds and near Great Ponton, south of Grantham. The Lincolnshire Limestone in Northamptonshire and Rutland, and the Great Oolite Limestone in Northamptonshire are important sources of gruond limestone and burnt lime for agriculture. The Lincolnshire Limestone is exploited for a limestone flux at South Witham, and at Little Bytham, near Bourne, the Great Oolite Limestone was dug for a similar purpose.

There are large quarries in the Carboniferous Limestone at Breedon, east of Burton-upon-Trent, where the rock is worked for limestone flux, road-stone, ground limestone for agriculture, and in the manufacture of burnt lime. At Lilleshall, near Newport, Shropshire, this formation is also worked for limestone flux. The Silurian limestones of Sedgley, Dudley, Walsall and Barr were worked for many centuries for lime and as a flux for ironstone, and at Much Wenlock the Wenlock Limestone is exploited as a source of road-stone, flux and agricultural lime.

### Iron Ore

The region includes parts of the outcrops of two of the most important sedimentary ironstone formations found in this country (Fig. 16). The lower horizon, the Marlstone Rock Bed, is usually a slightly ferruginous limestone, but in two areas it possesses the character of a calcareous ironstone. The more northerly field is centred on Melton Mowbray and includes parts of South Lincolnshire and Leicestershire, and the other is in the Banbury district. In both areas the ore consists of sideritic limestones and calcitic sideritic chamositic oolites. The ore is relatively free from detrital minerals. Generally the worked ironstone has been subjected to a process of secondary enrichment; weathering has caused the solution of calcium carbonate and oxidation of the iron minerals to limonite, which results in an increase in the iron content of the ore compared with the fresh rock. During 1967 the output from the Banbury Ironstone field was 313,300 long tons and from the South Lincolnshire and Leicestershire field 224,300 long tons.

The higher horizon, the important Northampton Sand Ironstone, extends southwards from the vicinity of Lincoln to near Towcester. The limits of this ironstone field are controlled by the lateral passage of the ironstone into ferruginous sandstone or its absence beneath younger strata. The formation contains several iron-rich rocks but the exploited ores usually consist of sideritic chamositic oolite and limonitic oolite. The main industrial development of the ironstone field commenced about 1852; in 1942 the ore output exceeded $10\frac{1}{2}$ million long tons and in 1967 production was 7,069,900 long tons. The main steel manufacturing towns are Corby, Kettering and Wellingborough. Almost all the iron ore produced is obtained by opencast methods, and, in the case of the Northampton Sand Ironstone, the overburden may be as much as 100 ft (Plate XI). In Staffordshire the ironstones of the Middle Coal Measures and the Blackband Ironstones of the Upper Coal Measures were formerly important as iron ores.

### Pottery Clays, Refractory Clays and Brick Clays

The Middle Coal Measures contain several excellent fireclays and refractory clays. They were used in the important pottery industry centred on Stoke-on-Trent but have been supplanted by Cornish china clays. They are also worked in South Staffordshire and Derbyshire for the manufacture of sanitary ware, drainpipes and pottery. In the past all the argillaceous formations and superficial deposits which crop out in Central England have been used in the local manufacture of bricks, tiles, etc. Today with increased mechanization within the industry and easier transport only the most suitable clays are used. The majority of the brickfields are sited on the Keuper Marl outcrop, and important concentrations of the industry occur in South Leicestershire and around Stafford. In the Staffordshire coalfields the Coal Measures mudstones and siltstones, particularly those of the Etruria Marl, are used in the manufacture of bricks and earthenware. At Stamford, fire-bricks, glazed pipes and other refractory wares are manufactured from the silts and clays of the Upper Estuarine Series. Around Peterborough, the Oxford Clay is extensively exploited as a brick clay. In Cheshire bricks and tiles are manufactured on a small scale from glacial till deposits.

**Sand and Gravel**

Much of the sand and gravel used for building purposes is obtained from glacial deposits which occur throughout the region. In places, for example along the Trent Valley, there are extensive workings in river terrace deposits. The relatively soft Permo-Triassic sandstones are exploited for building sands whilst the Bunter Pebble Beds are dug on a large scale in North Staffordshire for coarser material.

**Oil**

Seepages of oil have been recorded in two areas. In Coalbrookdale the Upper and Middle Coal Measures sandstones are found to be impregnated with tarry oil. Over a hundred years ago they were reputed to have yielded 1000 gallons of oil per week but despite borehole exploration the source rock has not been discovered. Similarly, exploration at Gun Hill, north of Leek, found only traces of oil though seepages are common in workings in the North Staffordshire Coalfield. The most promising result of the search for oil has been the discovery and development of the oilfield at Plungar between Nottingham and Grantham where the productive sands are of Lower Coal Measures and Millstone Grit Series age. The cumulative total crude oil production up to 30th June, 1968, is 34,114 long tons.

**Rock Salt and Brine**

Being soluble, salt cannot crop out at surface in Britain. Instead it has an 'incrop' against the base of the zone of ground-water circulation, which is generally between 200 and 400 ft (61–122 m) below the base of the drift deposits in Cheshire and Shropshire. The salt in the 'incrop' areas is overlain by marls which have been let down after solution of salt below, and these collapsed strata may also be brecciated. From ancient times men have sunk wells to the salt in these areas to win natural brine, once the main basis of the salt industry in this country, and the saltfields became notorious for the heavy surface damage caused by subsidence. Nowadays a large chemical industry is based on the controlled production of artificial brine below ground by a method which causes no subsidence, but salt for the table and for preserving is still largely produced from brine got by the old method. Where saliferous strata are too deep in the ground to be affected by ground-water no natural brine is formed at the top of the salt which is then said to have a 'dry rock-head'. 'Wet rock-head' conditions obtain in the brine areas, where the salt is nearer the surface. The old producing saltfields were confined to areas of 'wet rock-head', but the modern process of controlled brining can be operated whether the top of the salt is 'wet' or 'dry'.

The Meadowbank Mine, Winsford, Cheshire, is now the only working rock salt mine in the country. The rock salt is used for de-icing roads, and to a lesser extent for salt licks for cattle

Plate XI

The Mammoth W-1400 Dragline excavating deep Overburden in a Northampton Sand Ironstone pit

*(Photo: Stewarts and Lloyds Ltd.)*

*(For full explanation see p. ix)*

The bulk of the salt produced in Great Britain originates in the three saltfields within the region. Of these the largest is the Cheshire Saltfield (p. 71). The national salt production figure for 1966 is 7,217,000 tons, of which, 1,030,000 is rock salt and the remainder brine.

## Gypsum

The two thick beds of gypsum ($CaSO_4.2H_2O$, hydrated calcium sulphate) which occur in the higher part of the Keuper Marl are being extensively exploited for the manufacture of Portland Cement, special cements and Plaster of Paris. As mineral white it is used in the manufacture of paper, where it replaces kaolin as a filler. Impure gypsum is also used as a fertilizer (soil conditioner). The lower bed has been mined at East Leake in Leicestershire, at Chellaston near Derby, and at Fauld near Tutbury in Staffordshire. The close textured gypsum of Chellaston and Fauld has been worked for many centuries as ornamental alabaster. The upper beds are worked at Newark, and at several places between Nottingham and Leicester.

## Stone

In the past the quarrying of building stone was an important industry throughout the region, with the availability of suitable material being reflected in the character of the village architecture. In South Warwickshire, Northamptonshire, Leicestershire and Rutland the outcrop of the various divisions of the Jurassic can be traced by observing the stone used in local buildings. In contrast, in those areas which lacked suitable stone the houses were built with cobb or wattle and daub walls or with wood frames and brick. Hedges bound the fields in the clay areas whereas dry-stone walls are prevalent on the limestone uplands.

The Marlstone Rock Bed yields a beautiful golden brown freestone, the Hornton Stone, which has been worked at Edge Hill, near Banbury, since mediaeval times. The reddish brown Northampton Sand has also been extensively exploited. The Lincolnshire Limestone has yielded valuable building stone: several local names have been given to this stone, notably Ketton Stone, of which several Cambridge colleges are built; Weldon Stone, much of which is incorporated in the University Library building at Cambridge; and Clipsham Stone which has been used extensively for repair work on Canterbury Cathedral, York Minster, the Houses of Parliament and Oxford University; Barnack Stone, worked several centuries ago, was used in several cathedrals and abbeys and in local churches, for example in Peterborough Cathedral and Barnack Church; it was also used in some of the East Anglian castles. The Collyweston Slate has been quarried extensively for roofing tiles during the past 400 years, but at present only two shallow pits are still working. The 'slates' were worked either in open quarries or mined from small shafts. Blocks are laid out on the ground, with bedding planes vertical, and kept watered to prevent drying. Frost action enables the blocks to be split by hand into thin layers. It is found that if the blocks once become dry they lose their fissile character.

The varied green, grey and dark blue rocks of Charnwood Forest were used in building rugged but attractively coloured habitations, and the slates of Swithland and Woodhouse Eaves in attractive shades of dark blue and purple were widely used for roofing. Today only the Charnian plutonic rocks and the Mountsorrel Granodiorite are quarried for building material, almost all the output being used in the manufacture of artificial stone. Of the Triassic building stones those obtained from the Keuper Sandstone were most valued; they are brown, yellow or white in colour, and are often good freestones. Large disused quarries along the outcrop north and west of Wolverhampton (e.g. at Penkridge and Tong) indicate the extent to which the sandstone was formerly worked. It is still worked at the Hollington Quarries, north of Uttoxeter and at Grinshill, near Shrewsbury. The grits (sandstones) of the Millstone Grit Series have long been famous as grindstones and in recent years they have been used for this purpose in paper mills.

**Road-stones**

Rocks capable of withstanding the attrition produced by modern traffic are almost all of pre-Triassic age. In Charnwood Forest the 'syenites' of Groby, Markfield and Newhurst, the Peldar 'porphyroid' (Whitwick 'granite') and the 'porphyroids' at Bardon are being extensively exploited. The 'syenites' of Croft, Enderby and Narborough and the Mountsorrel Granodiorite are worked for road metal and also for kerb stones. The Cambrian quartzites of Hartshill and the Lickey Hills and the diorites intrusive in the Cambrian near Nuneaton are valuable sources of road chippings. The Carboniferous Limestone at Breedon is crushed for road metal, whilst the basalt interbedded in this formation at Doseley, Shropshire is also quarried. Basalts intrusive in the Coal Measures are quarried for both aggregate and 'cobbles' at Titterstone Clee Hill and near Rowley Regis. Small amounts of road metal suitable for minor roads are obtained from the Bunter Pebble Beds and from drift deposits.

**Ganisters, Refractory and Silica Sands**

The major sources of refractory sands are the Pleistocene deposits which are dug at Chelford, Cheshire. The Lower Estuarine Series is worked in the Kettering area and at Luffenham, Rutland. In the past, silica-bricks, furnace-bricks and ground ganister have been produced from the ganisters or highly siliceous seatearths of the Coal Measures.

**Moulding Sand**

Several of the sandstones of the Permo-Triassic are excellent moulding sands. Most of those exploited commercially form part of the Upper Mottled Sandstone formation. The foundation of the iron and brass foundries in the Birmingham area may have been associated with the abundance of moulding sands nearby. Sands are worked at Colesheath (Staffordshire), Wolverhampton, Bromsgrove and Kidderminster. The Pleistocene Congleton Sand of Cheshire is an increasingly important source of moulding sand and is exploited in large pits around Congleton.

## Water Supply

The region includes the major industrial centres of the Midlands and there is an ever increasing demand for water for both domestic and industrial use. Water from surface reservoirs and rivers satisfies a part of the demand. The relative lack of topographic relief, the high density of population and the relatively low rainfall have meant that there are few areas suitable for water storage. Several small reservoirs have been built in Northamptonshire, and on the Keuper Marl at the edges of Charnwood Forest. At Melbourne, Derbyshire, a dam has been constructed for a storage reservoir which stores water pumped from rivers in the Peak District of Derbyshire.

Supplies are obtained from underground sources throughout the region. The most important aquifers are the sandstones and pebble-beds of the Permo-Triassic. Water from the Bunter formation is commonly under an artesian head when tapped by a borehole from under a thick impervious cover of Keuper Marl. Water is also obtainable from the Keuper Sandstone, but where it is overlain by the Keuper Marl the water generally has a high permanent (non-carbonate) hardness. Water of this type is of great value in brewing, and numerous breweries have been established on the outcrop of the Keuper Marl, as at Burton-upon-Trent. Water from the Keuper beds is commonly saline, as for instance in a spring arising at the junction of the Keuper Sandstone and Keuper Marl at Leamington Spa, which is well known for its medicinal properties. The brine springs of Droitwich, known to the Romans, are still in active use for brine baths.

Local supplies are obtained from springs, and many of the villages on the outcrop of the Jurassic rocks use the water which issues from the base of the limestone formations or tap it through shallow wells. The limestones of the Lias only rarely yield water of potable quality, most of this being obtained from the Marlstone Rock Bed. In Northamptonshire and Rutland the Middle Jurassic rocks, notably the Northampton Sand and Lincolnshire Limestone, are important sources of underground water. Arenaceous glacial drift commonly contains large quantities of water and it is frequently tapped in Cheshire. Large supplies are obtained from river gravels, e.g. those of the Trent at Burton-upon-Trent, the Nene at Oundle and the Welland at Gretton near Uppingham.

# 12. Structure

The rocks of Central England have been affected by a long series of earth movements in which at least four major phases can be identified. These include the late Pre-Cambrian movements whose effects can be observed in the Nuneaton district, where the basal conglomerate of the Cambrian Hartshill Quartzite rests unconformably on sheared Pre-Cambrian igneous rocks. Elsewhere, as in Charnwood Forest, there is evidence of vulcanicity and orogenic movements in late Pre-Cambrian times. Also the Pre-Cambrian rocks are cleaved whereas this cleavage does not affect the Cambrian shales. The discovery of beds of Llanvirn age in a borehole at Great Paxton, north of St. Neot's, shows that the period of sedimentation which commenced with the deposition of the basal Cambrian quartzite continued into Ordovician times. This fixes more accurately the onset of the period of structural evolution during which the greater part of Central England formed a positive area lying between the Welsh sedimentary basin and that of Brabant.

Parts of the Midland counties may have constituted a relatively stable block at this period but some structural evolution may have accompanied the post-Tremadoc–pre-Wenlock phase of igneous activity which led to the emplacement of stocks and sills of diorite rocks in the Cambrian rocks. Silurian rocks extend from the Welsh Borders into the Birmingham area, but farther east near Atherstone, Warwickshire, the Upper Old Red Sandstone rests unconformably upon the Cambrian. Still farther east, in the South Derbyshire and Leicestershire coalfields, Carboniferous rocks rest directly on Cambrian or Pre-Cambrian, whilst in Charnwood Forest and certain areas to the south-east Mesozoic rocks, mainly of Triassic or Lower Jurassic age overlie the Pre-Cambrian basement. In these areas it is usually impossible to unravel the stages in the structural evolution in Palaeozoic times, but here, as in the areas farther west, there are indications that the ancient structures formed in the Pre-Cambrian rocks have exercised some degree of control over the orientation of younger structures.

The classification of the movements in terms of orogenic periods, e.g. Caledonian, Hercynian or Alpine (Tertiary) is therefore difficult and it is more satisfactory to classify them with reference to local stratigraphy, using such terms as Pre-Cambrian, pre-Old Red Sandstone, intra-Carboniferous, Permo-Carboniferous, post-Triassic and post-Jurassic.

The final stage in the structural evolution of the rocks forming the present land surface was the widespread development of large-scale superficial structures in Pleistocene times. They are due to mass movements in the rocks following the development of valleys cut by melt-water in ice and frozen ground and are most widely developed in Jurassic terrain where thick beds of silt and clay are present. Though their development may have been influenced by the initial dip of the strata or the presence of pre-existing fault and fissure patterns they are not themselves the direct result of deep-seated tectonic movement.

## Pre-Old Red Sandstone Movements

The limited area of outcrop of pre-Old Red Sandstone strata makes it impossible to describe in detail the pre-Old Red Sandstone structural evolution of Central England. The effects of the oldest tectonic movements represented in the region are seen in the Warwickshire Coalfield near Nuneaton and in Charnwood Forest. The Charnian rocks of the Forest were folded into a north-westerly trending anticlinorium with a south-easterly plunge during late Pre-Cambrian times. The intrusion of the 'syenites' of the Charnwood area, and the development of cleavage in the Charnian rocks were associated with these movements.

Lower Old Red Sandstone is present in Shropshire though it is overstepped locally by the Farlow Sandstone (Upper Old Red Sandstone) in the Titterstone Clee area. In the Warwickshire Coalfield, Upper Old Red Sandstone rests unconformably on Cambrian, while in the Wyboston Borehole in Bedfordshire, Upper Old Red Sandstone is underlain by Upper Devonian rocks which rest on Cambrian strata. Marine Upper Devonian and Upper Old Red Sandstone have been recorded in boreholes in Cambridgeshire and Northamptonshire. Thus there is an area in the south-east Midlands where the older Palaeozoic rocks, including Tremadoc, were folded and eroded prior to the deposition of a veneer of marine Upper Devonian sediments. These were succeeded by Upper Old Red Sandstone deposits marking a change of facies. Marine intercalations are found in the Warwickshire Coalfield where there is some evidence of growth of the Arley Fault in Old Red Sandstone times. The fault system bounding the east side of Titterstone Clee in Shropshire shows similar evidence of growth in Old Red Sandstone times. The main locus of the post-Cambrian pre-Carboniferous movements lay west of the Church Stretton Fault Complex where the Ordovician and Silurian rocks of Wales were strongly folded in the period preceding the deposition of the Upper Old Red Sandstone. The so-called Caledonian movements were completed by the close of Upper Old Red Sandstone–Lower Carboniferous times and the gentle south-westerly trending folds of the Brown Clee (Clee Hill) Syncline and Ludlow–Ledwyche Anticline date from this period. The Mountsorrel Granodiorite (p. 27) is thought to be an intrusion of post-Cambrian pre-Carboniferous age.

## Intra-Carboniferous and Permo-Carboniferous Movements

The present structural pattern of Central England, and in particular of the Midland coalfields, was largely evolved during the earth-movements which are generally grouped as the Hercynian orogeny. The intensive mountain building movements were mainly confined to the area south of the 'Armorican Front' (a line extending approximately from Dungeness to South Pembrokeshire). At this time Central England formed part of a relatively stable mass of Pre-Cambrian and Lower Palaeozoic rocks which may be described as the Midland Block. Consequently the main structural elements in the region take the form of faults, probably in many cases guided by pre-existing lines of weakness in the basement, with folding occupying a relatively minor role in the structural pattern.

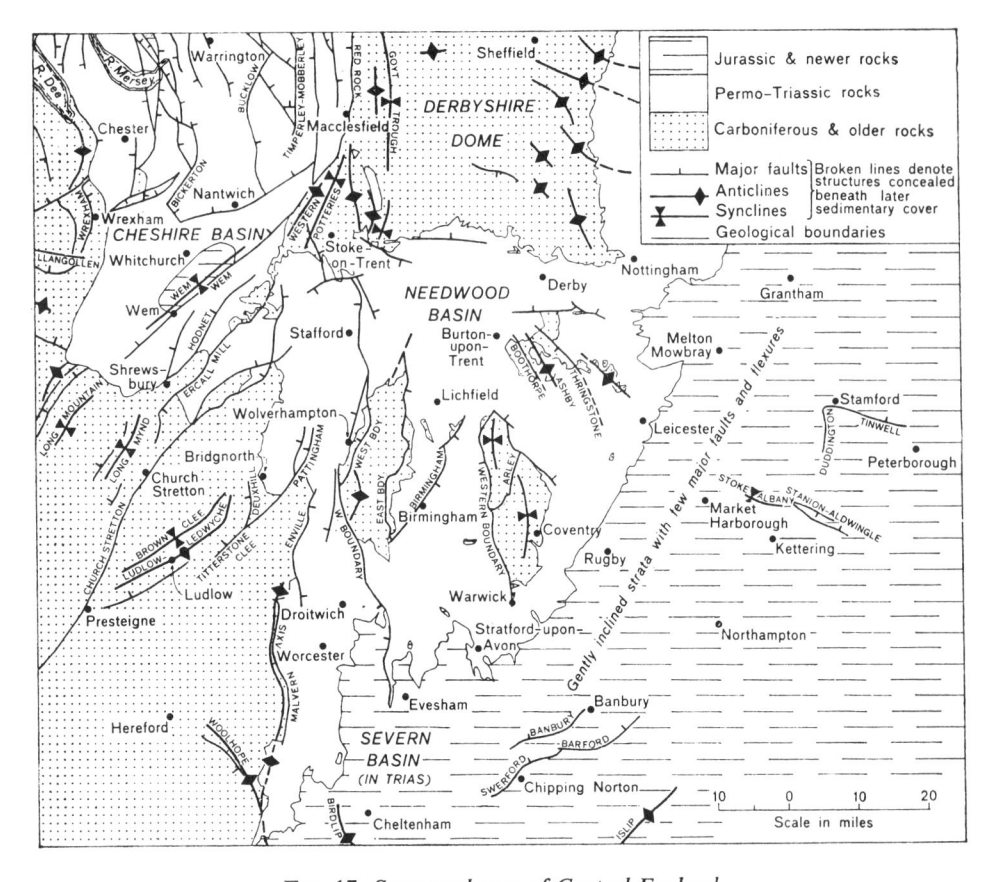

FIG. 17. *Structural map of Central England*

The effects of early Hercynian movements can be traced by the unconformities within the Carboniferous Limestone Series and Millstone Grit Series. In Coal Measures times breaks in deposition indicating differential subsidence are more marked. In some areas the Coal Measures increase in thickness in the troughs and thin over the crests of local folds; this is clearly seen in the Potteries Syncline and Western Anticline of the North Staffordshire (Potteries) Coalfield. It is evident that the folds were developing while the Coal Measures sediments were being laid down. Of the many local unconformities or non-sequences within the Coal Measures of Central England, the most striking is the 'Symon Fault' of the Coalbrookdale Coalfield (pp. 41, 48). Re-elevation of The Wrekin–Longmynd area led to erosion in pre-Keele times, while limestone breccias in the Keele Beds near Bridgnorth give evidence of the rise of a land area to the south.

The main movements took place at the end of the Carboniferous Period, and caused the transformation of the deltaic environment of Coal Measures times into the continental environment of the New Red Sandstone. The major north-south folds of the Pennine and Malvern uplifts were developed, together with relatively minor cross-folds such as that separating the

Leicester and Nottingham coalfields. At this time the main boundary faults of the Midland coalfields were also developed with the formation of the horst and graben structures mentioned earlier. The coalfields, now partly exposed, usually have a synclinal structure, seen most clearly in the Warwickshire Coalfield. The intervening graben areas acted as depositional troughs, and became infilled with later Permo-Triassic sediment.

## Post-Triassic Movements

The structural elements developed during the Permo-Carboniferous orogeny influenced the pattern of post-Triassic movements. These latter include local downwarping associated with minor folding, and strong faulting. The downwarping of certain areas, notably the Cheshire Basin, the Needwood Basin and the Severn Basin led to considerable modification of the early Mesozoic depositional basins while the post-Triassic faulting further accentuated the horst and graben structures of the Midlands coalfields and intervening areas. The margins of these horsts are generally well-defined topographically, since they commonly mark the boundary between the more resistant Palaeozoic rocks and the soft Triassic strata. Because of the evolution of these structures in Permo-Carboniferous and Triassic times the amount of post-Triassic movement along the faults cannot be determined. Exploratory boreholes in Cheshire have confirmed the importance and extent of this post-Triassic (possibly Tertiary) faulting. The Wem Fault which defines the eastern margin of the Prees Syncline has a throw of about 5000 ft (1500 m) and several faults with throws of at least 1000 ft (300 m) are present.

The numerous non-sequences within the Jurassic probably reflect in part the gentle movements of the Midland Block, but eustatic changes in sea-level may also have played a part in their formation. The most important disconformity is at the base of the Middle Jurassic. Traced to the south-east from the outcrop different formations of the Middle Jurassic rest upon various parts of the Upper and Middle Lias and beyond the region come to rest upon the Lower Lias in the area flanking the London Platform. In the same way the lower parts of the Upper Jurassic, the Upper Cornbrash and Kellaways Beds, rest non-sequentially upon different parts of the Middle Jurassic.

After the deposition of the Oxford Clay, slight movements continued, prior to the deposition of the Upper Cretaceous rocks now preserved in East Anglia and the London Basin. The tilting and very gentle flexuring movements which took place in post-Jurassic times probably reached their maximum during the mid-Tertiary or Alpine movements. The resulting easterly or south-easterly dip of the Jurassic rocks is remarkably uniform and is only locally modified by gentle folds. In places the Jurassic strata are affected by faults. Although these have throws of less than 200 ft (61 m) they are known to extend to the base of the Jurassic in places. They may reflect resumption of movement along pre-existing fault planes in the more rigid basement.

# Pleistocene Structures

## Structures due to mass movements

Structures due to large-scale mass movements of the rocks at or near the Pleistocene land surface are developed in many areas of Central England. They are generally described as superficial or non-diastrophic to distinguish them from structures produced solely by deep-seated tectonic movements. Mass movements of the superficial type are particularly well developed when the surface rocks consist of clays and silts of Jurassic age interbedded with limestone, sandstone or ironstone. Many of these structures are complex showing evidence of more than one period of growth. Nearly all of them have been dissected by late Pleistocene or Recent erosion and were formed before the younger river terraces and alluvial deposits were laid down. Where late Pleistocene or Recent erosion has been severe, as in some major river basins, only relics of these structures survive on higher ground. They attain their maximum development in those areas such as parts of the Middle Jurassic limestone plateaux where post-Glacial downcutting has been less severe.

The principal superficial structures may be classified as follows:

(a) *Cambers*, resulting from the lowering of outcropping or near-surface strata on the valley sides. The extent of the subsidence is largely determined by the extension and thinning of underlying clay and silt formations, the softened or waterlogged material having been removed by stream erosion. Subsidence of the surface was normally greatest on the lower slopes of the valleys. Thus the rocks acquired a valleyward tilt or camber producing a broad anticlinal structure on the inter-fluves.

(b) *Gulls* are widened, steeply inclined, downward tapering fissures, commonly aligned in closely spaced parallel belts along the flanks of major valleys but sometimes following pre-existing joint and fissure patterns. Many gulls contain infillings of Mesozoic formations or glacial deposits which collapsed into the fissures prior to the removal of the parent formations by sub-aerial erosion.

(c) *Dip-and-fault structure* is usually best developed where brittle rocks such as limestone or sandstone overlie clay or silt. The beds are broken into a series of faulted blocks or strips in which the rocks are tilted in a downhill direction at an angle which is normally somewhat steeper than the average dip. The effect of this tilting is to a large extent offset by a complementary upthrow of the fractured strata on the valleyward side of each fault. The downslope movement responsible for the formation of these structures may be regarded as a kind of 'creep' in which the strata have broken up in extensional movement. At the present time this kind of movement is restricted to shallow zones of weathering on slopes composed of clay, silt or marl. In Pleistocene times these structures were developed on a far greater scale and were commonly associated with heaving and collapse of the rocks in or below the valley floor. Dip-and-fault structures may be contrasted with step-faulting produced by rotational shear landslipping. The latter is associated with steep-sided valleys cut in soft rocks and is usually restricted to terminal slopes or over-deepened valleys in cambered areas. In such cases the faults hade to the downthrow, i.e. towards the valley floor and the step-faulted or landslipped formations may dip away from the valley floor and not towards it as in dip-and-fault structures.

Trough gulls may be associated with dip-and-fault structure due to the development of localized zones of tension in the broken material.

(d) *Sags* are areas or belts of strata where the rocks have collapsed or sunk following localized shearing or outflow of wet or supersaturated clay, sand and silt. They are usually found at the lower end of cambered valley slopes, adjacent to the central region of the valley.

(e) *Valley bulges*. In their simplest expression these are anticlinally folded masses of the material composing the valley floor. They include a number of structural types which may have resulted from differing structural processes. Some valley bulges are horst-like and are bounded by well-defined faults or sharp folds, as in some of the valleys of the Lincolnshire Limestone plateau (Hollingworth et al. 1944, fig. 7). Others are separated from the main camber slopes by sags. Some of the bulges in the Northampton Sand Ironstone Field appear to be located on older faults which, though primarily of tectonic origin, have been modified by later movements. Many valley bulges appear to have no obvious tectonic significance and some of the crumpled and contorted structures developed in the valleys cut in Lias clay may have been formed as a result of the development of glacial or post-glacial landslips (see below). Differential unloading of the valley floor by erosion and the drag effect produced by valleyward movement of a thinning ice cover may have played a part in the development of some of these structures. However, the growth of ground ice prior and subsequent to glaciation and the effects produced by thawing of perennially frozen ground are thought to be the most important process involved in their formation (Kellaway and Taylor 1953).

(f) *Landslips* due to rotational shear and chaotic hummocky slumps are present on some of the steeper slopes, notably on the outlier of Lias and Inferior Oolite at Bredon Hill, Worcestershire. Some of these slips postdate cambering of the Inferior Oolite. Landslips of recent origin give rise to uneven or hummocky slopes in which the relief of the slipped material has not been smoothed by erosion. Ancient landslips, possibly dating from the main phase of deglaciation of the Chalky Boulder Clay ice-sheet, are found in the valleys cut in the shales and sandstone of the Millstone Grit near Melbourne, Derbyshire. Here the surfaces of the landslipped material have been smoothed by erosion and the slips have no recognizable surface features. The shales proved beneath the slip planes in the valley floor are highly contorted and the structure is similar to that seen at Hollowell and the Eye Brook in the Lias of Northamptonshire (Hollingworth et al. 1944, figs. 13, 14). This suggests that similar rotational slips may also have been developed in Pleistocene times on the flanks of the deeper valleys incised in Lias Clay. Such structures would normally be very difficult to locate owing to the absence of surface features. They may well have been initiated when deglaciation was in progress and melt-water was tunnelling within or below the ice or when narrow, deeply incised valleys were being cut in ice and frozen clay rocks. With the development of the relief the process of thawing and erosion may have affected the stability of the soft Jurassic rocks. Then landslipping of the rocks composing the valley sides was accompanied by dragfolding and slight heaving of the clays and shales in the floor of the valleys. This explanation does not apply, however, to the horst-like or strongly folded valley bulges in which vertical movement predominated.

Most of the large Pleistocene structures of the Midlands can be related to mass movement taking place at an early stage in the development of the existing relief, probably during the period which includes the Chalky Boulder Clay glaciation. This ice-sheet covered much of the central and eastern Midlands. At its maximum development the glacial ice is likely to have been at least 750 feet in thickness, most of the high ground of Charnwood Forest having been buried. Ground which became perennially frozen in the period immediately preceding the arrival of this invading ice mass is likely to have thawed from below (Kellaway and Taylor 1953), once it was deeply buried beneath the glacial ice. If the base of the ice-sheet was at pressure-melting point considerable structural re-adjustment of the underlying rocks may have taken place owing to the load effect and the migration of water and supersaturated clay and silt in the underlying rocks.

Subsidence and faulting due to the melting of buried ground ice and folding due to the mobilization of clay and silt may therefore have been initiated during glaciation. In the later phases when ablation of the ice-sheet had proceeded to the stage when areas of bare ground were once more exposed to frost-action, or when the ice had thinned sufficiently to permit perennial freezing, further movements would be engendered by dissection of the surface ice and frozen ground by spring melt-water streams and by cyclic changes in precipitation leading to thickening and thinning of ice cover. Many sandy, silty and clay-rich rocks may have been rendered mobile by the passage of melt-water draining through the ground or trapped in taliks in frozen ground. The development of the large-scale superficial structures in the Midlands took place in stages extending over a considerable period of time and there is much local variation due to differing geological environments and to changes in the physical conditions in Pleistocene times. The fact that similar structures occur in the so-called periglacial area of southern England shows that the presence of a continental ice-sheet is not necessary to their formation though it may well have been responsible for the greater magnitude and variety of the Midland structures.

**Cryoturbation structures**

Large-scale superficial structures due to mass movements, are distinguishable from surface structures due to cryoturbation by their much greater size and regional effect. Very shallow disturbances due to cryoturbation such as involutions, fossil frost wedges, patterned ground and the like, produced at or near the surface of perennially or seasonally frozen ground, are also found in the Midlands. Some of these affect the terrace gravels or are developed in the frost-disintegrated material which mantles valley sides and which has accumulated as Head in the floors of dry valleys in the limestone uplands. For the most part these structures are younger than the cambers, bulges and other large-scale forms, and many of them were formed during the much later cold period represented by the younger terraces and 'buried channel' period (Table 2, p. 91).

# Quaternary Earth Movements
### Seismic activity and superficial deformation

The upper age limit of the faults cutting the Mesozoic rocks of the Midlands can seldom be determined with precision. In the east Midlands where Middle and Upper Jurassic rocks are present there are a number of important faults along which movement has taken place in post-Cornbrash or post-Oxford Clay times. Since there is no unequivocal evidence of the initiation of major faults in Quaternary times it appears that these structures are of Cretaceous or Tertiary age.

Shotton (1965) has suggested that faulting seen in glacial deposits at Kilsby, Northamptonshire, and Narborough in Leicestershire has resulted from the reactivation of faults in the basement during isostatic uplift following deglaciation. The structures at Kilsby and Narborough resemble trough gulls, and the trend of the bounding faults at Narborough appears to reflect the orientation of the faults in the Pre-Cambrian rocks of Charnwood Forest. The development of superficial structures in the Midlands has undoubtedly been influenced by the presence of pre-existing tectonic features. Yet there is no clear proof that reactivation of major structures has resulted from isostatic uplift. Seismic activity can, under favourable conditions, be induced in dormant or inactive faults by changes in loading (Carder 1945; McGinnis 1963). However, the field evidence suggests that glacial unloading and isostatic recovery in Central England is more likely to have been accompanied by a multiplicity of small internal adjustments than by large displacements along deep-seated faults.

Superficial manifestations of tectonic activity in seismically active regions include such structures as landslides, earth slumps, secondary fissures and earth lurches (Lawson 1908). The so-called fault block fissures formed in the Mississippi valley during the New Madrid earthquakes of 1811 and 1812 (Fuller 1912) have the same form as trough gulls, though the former develop on flat ground. Like trough gulls they owe their origin to tensional stresses due to lateral withdrawal of support. Somewhat similar structures are produced by the slow thawing of perennially frozen ground adjacent to lakes or rivers.

In Central England trough gulls are common on cambered valley sides or in the crests of cambered interfluves but are rare in undissected plateau areas. Gulls in feebly cambered plateau areas are generally parallel-sided and are rifts rather than troughs. This distribution supports the theory that gulls are caused primarily by lateral withdrawal of support, not purely vertical collapse as has sometimes been assumed. The observations made by Oldham (1899) on the structural deformation of alluviated valleys during the Assam Earthquakes of 1897 suggest that the formation of narrow strongly faulted valley bulges (Hollingworth *et al.* 1944) may in some degree have been assisted or triggered off by seismic activity. Bulging of the strata can be caused in several ways but the rocks in the valley floor would have been highly susceptible to rapid deformation at a time when melt-water was tunnelling in the base of the ice-sheet or cutting a deep narrow gorge in frozen ground. There is therefore every reason to consider the possible background influence of increased seismicity at the time that the Pleistocene structures were being formed.

Under present-day conditions the recorded earthquakes of Central England, which do not exceed intensity 8 on the Davison scale, are too weak to cause fissuring of the ground. The Colchester earthquake of 1884 is the only scientifically evaluated British earthquake to have attained intensity 9 (Davison 1924). On the other hand it may be that in Pleistocene times shocks of the same intensity as those of the present day had a considerable effect on partly frozen, ice-laden or waterlogged ground. Therefore a general increase in strength and frequency of shocks during the phase of isostatic uplift and dissection of the land surface might have been sufficient to trigger off extensive movements in the surface layers.

**Recent Earthquakes**

In historic times seismic activity in Central England has been slight and not all the tremors which have attracted public attention have originated in the region. It is to Davison (1924) that we owe most of our information. Summarized accounts of all the major shocks prior to January 1924 will be found in his book 'A History of British Earthquakes'.

Of those earthquakes which have their epicentres in Central England only a few can be related with any degree of certainty to known tectonic features. Thus the Leicester twin earthquake of 1893 may have been located on a north-north-westerly or north-westerly trending fault system (Charnian) passing through the Pre-Cambrian rocks of Charnwood Forest. Others, such as the Oakham earthquake of 1898 cannot be related to movement on any observed fault. The isoseismic axis of the Oakham disturbance has a north-westerly trend. It is paralleled on the surface by a strong belt of minor structures including many of Pleistocene age, notably between Oakham and Harringworth. Some of the old pre-glacial valleys may have a similar trend in this area. Possibly the Oakham earthquake originated on a north-westerly trending fault situated at depth in the basement. The minor structures in the Jurassic rocks may be a reflection of the later movements induced by feeble seismicity and superficial deformation in Pleistocene times.

The Stamford earthquakes of 1755, 1792, 1813 and 1844 are thought to have had a common epicentre situated about 3 miles south of Stamford on the Lincolnshire Limestone plateau near Wittering. This area is roughly bounded by the Duddington Fault on the west and the Tinwell Fault on the north. Both these faults cut the Oxford Clay. The Wittering plateau is underlain by a mass of ancient rocks, mainly Pre-Cambrian, over which the Mesozoic cover is thinner than in the adjacent areas. However, the predominant structural trend in mid-Jurassic times appears to have been S.W.–N.E. as indicated by the sedimentary belts formed on the margins of the evolving shorelines. The Duddington and Tinwell Faults cut across the structural and depositional trend of the Middle Jurassic rocks and are probably of post-Jurassic origin.

At the time when Davison (1924) compiled his account of the Stamford earthquakes only the Tinwell Fault was known. Since that time the area has been re-mapped in detail and is published on the one-inch Geological Survey Stamford (157) Sheet. The disturbed area of the Stamford earthquakes is roughly circular, that of the 1844 shock (probably the best documented one) being slightly elongated in a N.E.–S.W. direction. If this elongation may be taken as an indication of the trend of the structure responsible for the seismic activity it would appear unlikely that the 1844 shock was related to the Duddington or Tinwell faults. On the other hand there is a series of broad channels cut in the older Jurassic rocks. These have been infilled with Upper Lincolnshire Limestone and form a belt running in a N.E.–S.W. direction from Barnack to Kingscliffe. Similar channels occur in the Kettering area where Taylor (1947, 1963) recorded infilled channels of Upper Lincolnshire Limestone having the same trend. It seems probable that these channels were formed during the change from a marine to non-marine environment and that they mark a stage in the structural evolution of the shorelines in Jurassic times. The Stamford earthquake of 1844 may therefore have originated on an ancient basement structure with a Calcdonian trend which exercised control over the mid-Jurassic environments. In the Stamford earthquake of 1813 the longer dimension of the disturbed area was oriented in an east–west direction suggesting a relationship either with the Tinwell Fault as proposed by Davison or with parallel faults on the southern margin of the block. Thus the Stamford earthquakes are likely to have originated on a complex set of intersecting structures.

With regard to the general pattern of seismicity in Central England it may be observed that of the twelve earthquakes of intensity 5 or more (Davison scale) which have occurred in England since 1900, no less than three were located near Derby. It has been suggested by Davison (1924) that the twin earthquakes of Northampton (1750 and 1768), Leicester (1893 and 1904), Derby (1903–6) and Stafford (1916) may constitute a westerly progression. Westerly migration of foci is said to be characteristic of the Great Glen Fault in Scotland and is exemplified in the West of England by the Taunton and Barnstaple earthquakes of 1868–1920.

The Midland Earthquake of 1956 with its epicentre at Dizeworth near Derby (Dollar 1957) does not fit into a progression, though in the absence of detailed information about many minor shocks it is unwise to draw conclusions. This earthquake was one of the strongest yet recorded in Central England having an intensity of 8 on the Davison scale (7 on the modified Mercalli scale). A variety of damage was done to buildings and some chimneys were thrown down in Loughborough and Derby. Tillotson (*in* Dollar 1957) has calculated the probable depth of focus as 6–8 miles. The movement may have originated on a north-westerly trending structure but there is insufficient evidence to relate it to any particular fault whose surface position lies near the epicentre. In general the problem of relating recent earthquakes to deeply buried structures is difficult and up to the present time little progress has been made in respect of Central England.

# 13. Geological Survey Maps and Memoirs, and Other References Relevant to Central England

## Maps

(a) **On the Scale of 4 miles to 1 in (1/253,440)**
*Colour-printed; Solid Edition (except Sheet 12); Out of Print (except Sheet 12).*
Sheet 9 (with 10). Chester, Shrewsbury.
" 11. Derby, Lincoln, Stafford.
" 12. Louth, Peterborough, Norwich.
" 14. Hereford, Clee Hills.
" 15. Birmingham, Northampton, Oxford, Worcester.

(b) **On the Scale of 1 mile to 1 in (1/63,360) or 1/50,000**
The Old Series Sheets are hand coloured and are not now in print (those replaced by New Series Sheets are obsolete and are not listed). The New Series Sheets are colour printed; some are printed in separate 'solid' and 'drift' editions, others in one 'solid and drift' or 'solid with drift' edition, and others only in a 'drift' edition.

(i) *New Series Sheets*
109 (Chester); 110 (Macclesfield); 121 (Wrexham); 122 (Nantwich); 123 (Stoke upon Trent), out of print; 125 (Derby); 126 (Nottingham); 137 (Oswestry); 138 (Wem); 139 (Stafford); 140 (Burton upon Trent); 141 (Loughborough); 142 (Melton Mowbray); 143 (Bourne); 152 (Shrewsbury); 153 (Wolverhampton); 154 (Lichfield); 155 (Atherstone); 156 (Leicester); 157 (Stamford); 166 (Church Stretton); 167 (Dudley); 168 (Birmingham); 169 (Coventry); 170 (Market Harborough); 171 (Kettering); 182 (Droitwich); 185 (Northampton); 186 (Wellingborough); 200 (Stratford-upon-Avon); 201 (Banbury); 202 (Towcester); 217 (Moreton in Marsh); 218 (Chipping Norton)

(ii) *Old Series Sheets*
For the areas where no New Series Sheets have been published the following Old Series Sheets should be consulted:
44 (Evesham, Tewkesbury, Cheltenham, Burford); 45 N.W. (Banbury, Deddington, Chipping Norton); 45 N.E. (Buckingham, Brackley); 52 N.W.* (Kettering, Wellingborough, Thrapston); 52 S.W. (Northampton, Olney, Harrold); 53 N.W. (Coventry, Rugby, Leamington); 53 N.E.* (Clipston, Crick, Braunston); 53 S.W.* (Southam, Kineton); 53 S.E. (Towcester, Daventry, Weedon); 54 N.W.* (Droitwich, Bromsgrove); 54 N.E. (Henley in Arden, Solihull); 54 S.W. (Worcester, Pershore); 54 S.E. (Stratford upon Avon, Alcester); 55 N.W. (Titterstone Clee Hill, Ludlow); 55 N.E.* (Cleobury Mortimer, Bewdley, Stourport, Kidderminster); 61 S.W.* (Church Stretton, Brown Clee Hill); 61 S.E.* (Much Wenlock, Bridgnorth); 63 S.E. (Lutterworth, Market Harborough); 64 (Melton Mowbray, Oakham, Uppingham, Stamford, Peterborough); 70 (Newark, Grantham, Corby, Sleaford, Spalding, Tattersall); 72 N.W.* (Hanley, Stoke upon Trent, Newcastle under Lyme, Cheadle); 72 N.E. (Ashbourne, Dovedale); 72 S.E.* (Burton upon Trent, Tutbury, Uttoxeter); 81 S.W.* (Macclesfield, Congleton)

*Largely replaced by New Series Sheets.

(c) **On the Scale of about 2½ in to 1 mile (1/25,000)**
SO48 (Craven Arms)
SO49 (Church Stretton)
SO59 (Wenlock Edge)
SP83 with parts of SP73, 74, 84, 93, and 94 (Milton Keynes)
Parts of TF00, 10, 20, TL09, 19, 29 (Peterborough)

(d) **On the Scale of 6 in to 1 mile (1/10,560)**
The greater part of the area represented by New Series one-inch scale maps is also covered by maps on the six-inch scale. For most of the areas where Coal Measures are exposed the maps are published. The six-inch maps of the other areas are deposited for reference in the library of the Institute of Geological Sciences, Exhibition Road, London, S.W.7, and at the Institute of Geological Sciences, Ring Road Halton, Leeds 15, where they may be consulted. Uncoloured photo-copies may be supplied on special order.

(e) **Geophysical Maps**
(i) **On the Scale of 4 miles to 1 in (1/253,440) Gravity Survey Overlay**
Sheet 11 (1956). Stockport, Lincoln, Wolverhampton, Stamford
,, 12 (1960). Louth, Peterborough, Norwich
,, 15 (1954). Birmingham, Northampton, Gloucester, Oxford, Worcester
(ii) **National Grid Diagram Edition (1/250,000) Aeromagnetic**
Sheet 5. English Midlands and Welsh Borders

(f) **On the scale of ½ in to 1 mile (1/126,720)**
Hydrogeological Map of North and East Lincolnshire.

## Memoirs

(a) **General Memoirs**
BARROW, G. 1903. The Geology of the Cheadle Coalfield.
CROOKALL, R. 1955–. Fossil Plants of the Carboniferous Rocks of Great Britain [Second Section]. *Mem. Geol. Surv. Palaeont.*, **4**
FOX-STRANGWAYS, C. and WOODWARD, B. 1892–5. The Jurassic Rocks of Britain. 5 vols. (Jurassic rocks of Central England are referred to in vols. iii, iv and v.)
GIBSON, W. 1905. The Geology of the North Staffordshire Coalfields.
HOWELL, H. H. 1859. The Geology of the Warwickshire Coalfield and the Permian Rocks and Trias of the surrounding district.
HULL, E. 1860. The Geology of the Leicestershire Coalfield and of the Country around Ashby-de-la-Zouch.
────── 1869. The Triassic and Permian Rocks of the Midland Counties of England.
JUDD, J. W. 1875. The Geology of Rutland.
JUKES, J. B. 1859. The South Staffordshire Coalfield. 2nd edit.
KIDSTON, R. 1923–5. Fossil Plants of the Carboniferous Rocks of Great Britain [First Section]. *Mem. Geol. Surv. Palaeont.*, **2**, Pts. 1–6.
LAMPLUGH, G. W. and GIBSON, W. 1910. The Geology of the Country around Nottingham.
MITCHELL, G. H. and STUBBLEFIELD, C. J. 1948. The Geology of the Leicestershire and South Derbyshire Coalfield. 2nd edit. *Geol. Surv. Wartime Pamphlet*, **22**.
────── ────── and CROOKALL, R. 1942. The Geology of the Warwickshire Coalfield. *Geol. Surv. Wartime Pamphlet*, **25**.
────── ────── ────── 1945. The Geology of the Northern Part of the South Staffordshire Coalfield (Cannock Chase Region). *Geol. Surv. Wartime Pamphlet*, **43**.
SKERTCHLY, S. B. J. 1877. The Geology of the Fenland.

WHITEHEAD, T. H. and EASTWOOD, T. 1927. Geology of the Southern Part of the South Staffordshire Coalfield.

(b) **Sheet Memoirs**

Memoirs descriptive of most of the one-inch geological maps have been published. Those referring to Old Series Sheets are now out of print and are not listed. Those referring to New Series Sheets are as follows:

BARROW, G., GIBSON, W., CANTRILL, T. C., DIXON, E. E. L. and CUNNINGTON, C. H. 1919. The Geology of the Country around Lichfield (Sheet 154).

EASTWOOD, T., GIBSON, W., CANTRILL, T. C. and WHITEHEAD, T. H. 1923. The Geology of the Country around Coventry (Sheet 169).

———, WHITEHEAD, T. H. and ROBERTSON, T. 1925. The Geology of the Country around Birmingham (Sheet 168).

EDMONDS, E. A., POOLE, E. G. and WILSON, V. 1965. Geology of the Country around Banbury and Edge Hill (Sheet 201).

EVANS, W. B., WILSON, A. A., TAYLOR, B. J. and PRICE, D. J. 1968. The Geology of the Country around Macclesfield, Congleton, Crewe and Middlewich (Sheet 110).

FOX-STRANGWAYS, C. 1903. The Geology of the Country near Leicester (Sheet 156).

——— and WATTS, W. W. 1900. The Geology of the Country between Atherstone and Charnwood Forest (Sheet 155).

——— ——— 1905. The Geology of the Country between Derby, Burton-on-Trent, Ashby-de-la-Zouch and Loughborough (Sheet 141).

GIBSON, W. 1925. The Geology of the Country around Stoke-upon-Trent (Sheet 123).

——— POCOCK, T. I., WEDD, C. B. and SHERLOCK, R. L. 1908. The Geology of the southern part of the Derbyshire and Nottinghamshire Coalfield (Sheet 125).

GREIG, D. C., WRIGHT, J. E., HAINS, B. A. and MITCHELL, G. H. 1968. Geology of the Country around Church Stretton, Craven Arms, Wenlock Edge and Brown Clee (Sheet 166).

LAMPLUGH, G. W., GIBSON, W., SHERLOCK, R. L. and WRIGHT, W. B. 1908. The Geology of the Country between Newark and Nottingham (Sheet 126).

———  ———, WEDD, C. B., SHERLOCK, R. L. and SMITH, B. 1909. The Geology of the Melton Mowbray district and south-east Nottinghamshire (Sheet 142).

MITCHELL, G. H., POCOCK, R. W. and TAYLOR, J. H. 1962. Geology of the Country around Droitwich, Abberley and Kidderminster (Sheet 182).

POCOCK, R. W., WHITEHEAD, T. H., WEDD, C. B. and ROBERTSON, T. 1938. Shrewsbury District (including the Hanwood Coalfield) (Sheet 152).

——— and WRAY, D. A. 1925. The Geology of the Country around Wem (Sheet 138).

POOLE, E. G. and WHITEMAN, A. J. 1966. Geology of the Country around Nantwich and Whitchurch (Sheet 122).

———, WILLIAMS, B. J. and HAINS, B. A. 1968. The Geology of the Country around Market Harborough (Sheet 170).

RICHARDSON, L. 1929. The Country around Moreton in Marsh (Sheet 217).

STEVENSON, I. P. and MITCHELL, G. H. 1955. Geology of the Country between Burton upon Trent, Rugeley and Uttoxeter (Sheet 140).

TAYLOR, J. H. 1963. Geology of the Country around Kettering, Corby and Oundle (Sheet 171).

WEDD, C. B., SMITH, B., KING, W. B. R. and WRAY, D. A. 1929. The Country around Oswestry (Sheet 137).

WEDD, C. B., SMITH, B. and WILLS, L. J. 1927-8. The Geology of the Country around Wrexham (Sheet 121). Pt. I (1927). Lower Palaeozoic and Lower Carboniferous Rocks. Pt. II (1928). Coal Measures and Newer Formations.

WHITEHEAD, T. H., DIXON, E. E. L., POCOCK, R. W., ROBERTSON, T. and CANTRILL, T. C. 1927. The Country between Stafford and Market Drayton (Sheet 139).

―――― and POCOCK, R. W. 1947. Dudley and Bridgnorth (Sheet 167).

――――, ROBERTSON, T., POCOCK, R. W. and DIXON, E. E. L. 1928. The Country between Wolverhampton and Oakengates (Sheet 153).

WILLIAMS, B. J. and WHITTAKER, A. 1974. Geology of the country around Stratford-upon-Avon and Evesham (Sheet 200).

(c) **Economic Memoirs**
   (i) *Water Supply*
   BUTLER, A. J. and LEE, J. 1943. Water Supply from Underground Sources of the Birmingham–Gloucester District (Quarter-Inch Geological Sheet 15, Western Half). *Geol. Surv. Wartime Pamphlet*, **32.**
   
   LAMPLUGH, G. W. and SMITH, B. 1914. The Water Supply of Nottinghamshire from underground sources.
   
   LAND, D. H. 1966. Hydrogeology of the Bunter Sandstone in Nottinghamshire. *Water Supply Papers, Geol. Surv. Gt. Brit., Hydrogeological Rpt.* **1.**
   
   ―――― 1966. Hydrogeology of the Triassic Sandstones in the Birmingham–Lichfield district. *Water Supply Papers, Geol. Surv. Gt. Brit., Hydrogeological Rpt.* **2.**
   
   RICHARDSON, L. 1928. Wells and Springs of Warwickshire.
   
   ―――― 1930. Wells and Springs of Worcestershire.
   
   ―――― 1931. Wells and Springs of Leicestershire.
   
   STEPHENS, J. V. 1929. Wells and Springs of Derbyshire.
   
   WHITAKER, W. 1922. The Water Supply of Cambridgeshire, Huntingdonshire and Rutland from underground sources.
   
   WOODLAND, A. W. 1940-3. Water Supply from Underground Sources of the Oxford–Northampton District (Quarter-Inch Geological Sheet 15, Eastern Half). *Geol. Surv. Wartime Pamphlet*, **4.**

   (ii) *Mineral Resources*
   DEWEY, H. and EASTWOOD, T. 1925. Copper Ores of the Midlands, Wales, The Lake District and the Isle of Man. *Mem. Geol. Surv., Min. Resources*, **30.**
   
   HOLLINGWORTH, S. E. and TAYLOR, J. H. 1951. The Mesozoic Ironstones of England. The Northampton Sand Ironstone: Stratigraphy, Structure and Reserves.
   
   HOWE, J. A. (Editor). 1920a. Refractory Materials: Ganister and Silica Rock-Sand for Open-Hearth Steel Furnaces–Dolomite, 2nd edit. *Mem. Geol. Surv., Min. Resources*, **6.**
   
   ―――― 1920b. Refractory Materials: Fireclays. *Mem. Geol. Surv., Min. Resources*, **14.**
   
   LAMPLUGH, G. W., WEDD, C. B. and PRINGLE, J. 1920. Iron Ores: Bedded Ores of the Lias, Oolites and Later Formations in England. *Mem. Geol. Surv., Min. Resources*, **12.**
   
   SHERLOCK, R. L. 1921. Rock-Salt and Brine. *Mem. Geol. Surv., Min. Resources*, **18.**
   
   ―――― and HOLLINGWORTH, S. E. 1938. Gypsum and Anhydrite, and Celestine and Strontianite. 3rd edit. *Mem. Geol. Surv., Min. Resources*, **3.**
   
   STRAHAN, A. and OTHERS. 1920. Iron Ores: Pre-Carboniferous and Carboniferous Bedded Ores of England and Wales. *Mem. Geol. Surv., Min. Resources*, **13.**

TAYLOR, J. H. 1949. The Mesozoic Ironstones of England. The Petrology of the Northampton Sand Ironstone Formation.
WHITEHEAD, T. H., ANDERSON, W., WILSON, V. and WRAY, D. A. 1952. The Mesozoic Ironstones of England. The Liassic Ironstones.

(d) **Explanations of Maps on the Scale of about 2½ in to 1 mile (1/25,000)**
HAINS, B. A. 1969. The geology of the Craven Arms area (Sheet SO48).
────── 1970. The geology of the Wenlock Edge area (Sheet SO59).
HORTON, A., LAKE, R. D., BISSON, G. and COPPACK, B. C. 1974. The geology of Peterborough (Special Sheet including parts of TF00, 10, 20, TL09, 19, 29).
────── SHEPHARD-THORN E. R. and THURRELL, R. G. 1974. The geology of the new town of Milton Keynes (Special Geological Sheet SP83 with parts of SP73, 74, 84, 93 and 94).
WRIGHT, J. E. 1968. The geology of the Church Stretton area (Sheet SO49).

## List of other References
### General
BRODIE, P. B. 1850. Sketch of the geology of the neighbourhood of Grantham, Lincolnshire ... *Ann. Mag. Nat. Hist.*, **6**, 256–66.
DINELEY, D. L. 1960. Shropshire Geology: An Outline of the Tectonic History. *Field Studies*, **1**, 86–108.
FALCON, N. L. and KENT, P. E. 1960. Geological Results of Petroleum Exploration in Britain 1945–1957. *Mem. Geol. Soc., London*, No. 2.
FAREY, J. 1808. *Horizontal Geological Sections*. Manuscript in Geological Survey and Museum Library, London.
GARRETT, P. A., HARDIE, W. G., LAWSON, J. D. and SHOTTON, F. W. 1958. Geology of the area around Birmingham. *Geol. Assoc. Guide*, **1**.
LAPWORTH, C., WATTS, W. W. and HARRISON, W. J. 1898. Sketch of the Geology of the Birmingham District. *Proc. Geol. Assoc.*, **15**, 313–416.
LEES, G. M. and TAITT, A. H. 1946. The geological results of the search for oilfields in Great Britain. *Quart. J. Geol. Soc.*, **101** (for 1945), 255–317.
MARSHALL, C. E. and OTHERS. 1948. *Guide to the Geology of the East Midlands*. Nottingham.
MENEISY, M. Y. and MILLER, J. A. 1963. A Geochronological Study of the Crystalline Rocks of Charnwood Forest, England. *Geol. Mag.*, **100**, 507–23.
MITCHELL, G. H., with appendices by SABINE, P. A. and PONSFORD, D. R. A. 1954. The Whittington Heath Borehole. *Bull. Geol. Surv. Gt. Brit.*, **5**, 1–60.
MURCHISON, R. I. 1839. *Silurian System*. London.
PLOT, R. 1686. *Natural History of Staffordshire*. Oxford.
RASTALL, R. H. 1925. On the Tectonics of the Southern Midlands. *Geol. Mag.*, **62**, 193–222.
SMITH, B. 1913. The Geology of the Nottingham District. *Proc. Geol. Assoc.*, **24**, 205–40.
SMITH, W. 1815. *Strata of England and Wales*. Memoir to the map and delineation of the strata of England and Wales with part of Scotland. London.
TRUEMAN, A. E. (Editor). 1954. *The Coalfields of Great Britain*. London.
VARIOUS AUTHORS. 1910. Geology in the Field: The Jubilee Volume of the Geologists' Association (Northamptonshire, Rutland and Warwickshire, 450–87; Nottinghamshire, 518–39; Staffordshire, 564–91; Shropshire, 739–69; Charnwood Forest, 770–85).
WATTS, W. W. 1925. The Geology of South Shropshire. *Proc. Geol. Assoc.*, **36**, 321–63.
────── 1947. *Geology of the Ancient Rocks of Charnwood Forest*. Leicester.
WHITTARD, W. F. 1952. A Geology of South Shropshire. *Proc. Geol. Assoc.*, **63**, 143–97.

WILLS, L. J. 1935. An outline of the Palaeogeography of the Birmingham Country. *Proc. Geol. Assoc.*, **46**, 211–46.
——— 1950. *The Palaeogeography of the Midlands*. 2nd edit., Liverpool.
——— 1951. *A Palaeogeographical Atlas of the British Isles and Adjacent Parts of Europe*. London and Glasgow.
——— 1956. *Concealed Coalfields*. London and Glasgow.
YATES, J. 1829. Observations on the structure of the border country of Salop and North Wales; and of some detached groups of transition rocks in the Midland Counties. *Trans. Geol. Soc.* (2), **2**, 237–64.

### Pre-Cambrian

ALLEN, J. R. L. 1957. The Pre-Cambrian Geology of Caldecote and Hartshill, Warwickshire. *Trans. Leicester lit. phil. Soc.*, **51**, 16–31.
BENNETT, F. W., LOWE, E. E., GREGORY, H. H. and JONES, F. 1928. The Geology of Charnwood Forest. *Proc. Geol. Assoc.*, **39**, 241–98.
EVANS, A. M. 1963. Conical Folding and Oblique Structures in Charnwood Forest, Leicestershire. *Proc. Yorks. Geol. Soc.*, **34**, 67–80.
FORD, T. D. 1958. Pre-Cambrian Fossils from Charnwood Forest. *Proc. Yorks. Geol. Soc.*, **31**, 211–7.
WILLS, L. J. and SHOTTON, F. W. 1934. New Sections showing the Junction of the Cambrian and Pre-Cambrian at Nuneaton. *Geol. Mag.*, **71**, 512–21.

### Cambrian

BUTTERLEY, A. D. and MITCHELL, G. H. 1945. Driving of Two Drifts by the Desford Coal Co. Ltd., at Merry Lees, Leicestershire. *Trans. Inst. Mining Eng.*, **104**, 703–13.
ILLING, V. C. 1913. Recent Discoveries in the Stockingford Shales near Nuneaton. *Geol. Mag.* (5), **10**, 452–3.
——— 1916. The Paradoxidian Fauna of a part of the Stockingford Shales. *Quart. J. Geol. Soc.*, **71** (for 1915), 386–450.
LAPWORTH, C. 1886. On the Sequence and Systematic Position of the Cambrian Rocks of Nuneaton. *Geol. Mag.* (3), **3**, 319–22.
RUSHTON, A. W. A. 1966A. In *Sum. Prog. Geol. Surv.* for 1965, 69.
——— 1966B. A monograph of the Cambrian trilobites from the Purley Shales of Warwickshire. *Palaeont. Soc.*
SMITH, J. D. D. and WHITE, D. E. 1963. Cambrian Trilobites from the Purley Shales of Warwickshire. *Palaeontology*, **6**, 397–407.

### Silurian and Old Red Sandstone

ALLEN, J. R. L. and TARLO, L. B. 1963. The Downtonian and Dittonian Facies of the Welsh Borderland. *Geol. Mag.*, **100**, 129–55.
BALL, H. W. 1951. The Silurian and Devonian Rocks of Turner's Hill and Gornal, South Staffordshire. *Proc. Geol. Assoc.*, **62**, 225–36.
———, DINELEY, D. L. and WHITE, E. I. 1961. The Old Red Sandstone of Brown Clee Hill and the adjacent area. *Bull. Brit. Mus. (Nat. Hist.) Geol.*, **5**, 175–310.
BUTLER, A. J. 1937. On Silurian and Cambrian rocks encountered in a Deep Boring at Walsall, South Staffordshire. *Geol. Mag.*, **74**, 241–57.
——— 1939. The Stratigraphy of the Wenlock Limestone of Dudley. *Quart. J. Geol. Soc.*, **95**, 37–74.
KING, W. W. 1921. The Plexography of South Staffordshire in Avonian Time. *Trans. Inst. Mining Eng.*, **61**, 151–68.
——— 1924. The Geology of Trimpley. *Trans. Worcs. Nat. Cl.*, **7**, 319–22.

KING, W. W. 1925. Notes on the "Old Red Sandstone" of Shropshire. *Proc. Geol. Assoc.*, **36**, 383–9.

―――― 1934. The Downtonian and Dittonian Strata of Great Britain and North-Western Europe. *Quart. J. Geol. Soc.*, **90**, 526–66.

LOWE, E. E. 1926. *The Igneous Rocks of the Mountsorrel District*. Leicester.

MILLER, J. A. and PODMORE, J. S. 1961. Age of Mountsorrel Granite. *Geol. Mag.*, **98**, 86–8.

SQUIRRELL, H. C. 1958. New Occurrences of Fish Remains in the Silurian of the Welsh Borderland. *Geol. Mag.*, **95**, 328–32.

TARLO, L. B. H. 1964. Psammosteiformes (Agnatha)—A review with descriptions of new material from the Lower Devonian of Poland. I. General Part. *Palaeont. Pol.*, **13**, 1–135.

TAYLOR, K. 1966. In *Sum. Prog. Geol. Surv.* for 1965, 47.

WHITE, E. I. 1946. The Genus *Phialaspis* and the "*Psammosteus* Limestones". *Quart. J. Geol. Soc.*, **101** (for 1945), 207–42.

―――― 1950. The Vertebrate Faunas of the Lower Old Red Sandstone of the Welsh Borders. *Bull. Brit. Mus. (Nat. Hist.) Geol.*, **1**, 51–67.

WILLS, L. J. in collaboration with WILKINS, L. G. and HUBBARD, G. H. 1925. "New Exposures in the Rubery–Longbridge–Rednal District, south of Birmingham". B. The Upper Llandovery Series of Rubery. *Proc. Birm. Nat. Hist. and Phil. Soc.*, **15**, 67–83.

**Carboniferous**

CROOKALL, R. 1931. A Critical Revision of Kidston's Coal Measure Floras. *Proc. Roy. Phys. Soc. Edin.*, **22**, 1–34.

―――― 1955. Fossil Plants of the Carboniferous Rocks of Great Britain. [Second Section]. *Mem. Geol. Surv. Palaeont.*, **4**, pt. 1.

EDWARDS, W. and TROTTER, F. M. 1954. The Pennines and adjacent areas. 3rd edit. *Brit. Reg. Geol.*

EUNSON, H. J. 1884. The Range of the Palaeozoic Rocks beneath Northampton. *Quart. J. Geol. Soc.*, **40**, 482–96.

EVERITT, C. W. F. 1960. Rock Magnetism and the Origin of the Midland Basalts. *Geophys. J. R. astr. Soc.*, **3**, 203–10.

GEORGE, T. N. 1956. The Namurian Usk Anticline. *Proc. Geol. Assoc.*, **66**, 297–316.

―――― 1958. Lower Carboniferous Palaeogeography of the British Isles. *Proc. Yorks. Geol. Soc.*, **31**, 227–318.

GIBSON, W. 1913. A Boring for Coal at Claverley, near Bridgnorth, and its bearing on the extension westwards of the South Staffordshire Coalfield. *Trans. Inst. Mining Eng.*, **45**, 30–48.

HIND, W. 1894–6. A Monograph on *Carbonicola, Anthracomya* and *Naiadites*. *Palaeontogr. Soc.*

HOARE, R. H. 1959. Red Beds in the Coal Measures of the West Midlands. *Trans. Inst. Mining Eng.*, **119**, 185–98.

HORTON, A. 1963. In *Sum. Prog. Geol. Surv.* for 1962, 39.

JONES, D. G. and OWEN, T. R. 1961. The Age and Relationships of the Cornbrook Sandstone. *Geol. Mag.*, **98**, 285–91.

JONGMANS, W. J. 1928. Discussion générale. *C. R. Cong. Strat. Carb. Heerlen, 1927*, xlii–xliv.

KIDSTON, R. 1894. On the Various Divisions of British Carboniferous Rocks as determined by their Fossil Flora. *Proc. Roy. Phys. Soc. Edin.*, **12**, 183–257.

―――― 1905. On the Divisions and Correlation of the Upper Portion of the Coal-Measures, with special reference to their Development in the Midland Counties of England. *Quart. J. Geol. Soc.*, **61**, 308–23.

KIDSTON, R. 1923–5. Fossil Plants of the Carboniferous Rocks of Great Britain. [First Section]. *Mem. Geol. Surv., Palaeont.*, **2**, Pts. 1–6.

———, CANTRILL, T. C. and DIXON, E. E. L. 1917. The Forest of Wyre and the Titterstone Clee Coalfields. *Trans. Roy. Soc. Edin.*, **51**, 999–1084.

MARSHALL, C. E. 1942. Field Relations of certain of the basic igneous rocks associated with the Carboniferous Strata of the Midland counties. *Quart. J. Geol. Soc.*, **98**, 1–25.

MITCHELL, G. H. and STUBBLEFIELD, C. J. 1941. The Carboniferous Limestone of Breedon Cloud, Leicestershire, and the associated inliers. *Geol. Mag.*, **78**, 201–19.

POCOCK, R. W. 1926. The Basalt of Little Wenlock (Shropshire). *Sum. Prog. Geol. Surv.* (for 1925), 140–56.

——— 1931. The Age of the Midland Basalts. *Quart. J. Geol. Soc.*, **87**, 1–12.

SPINK, K. 1965. Coalfield Geology of Leicestershire and South Derbyshire: The Exposed Coalfields. *Trans. Leicester lit. phil. Soc.*, **59**, 41–98.

STEPHENS, J. V., EDWARDS, W., STUBBLEFIELD, C. J. and MITCHELL, G. H. 1942. The Faunal Divisions of the Millstone Grit Series of Rombalds Moor and neighbourhood. *Proc. Yorks. Geol. Soc.*, **24**, 344–72.

STUBBLEFIELD, C. J. and TROTTER, F. M. 1957. Divisions of the Coal Measures on Geological Survey Maps of England and Wales. *Bull. Geol. Surv. Gt. Brit.*, No. 13, 1–5.

TRUEMAN, A. E. and WEIR, J. 1946–68. A Monograph of British Carboniferous Non-Marine Lamellibranchia. *Palaeontogr. Soc.*

VAUGHAN, A. 1905. The Palaeontological Sequence in the Carboniferous Limestone of the Bristol Area. *Quart. J. Geol. Soc.*, **61**, 181–307.

**Enville Beds**

BOULTON, W. S. 1924. On a recently discovered breccia-bed underlying Nechells (Birmingham). *Quart. J. Geol. Soc.*, **80**, 343–73.

——— 1933. The Rocks between the Carboniferous and the Trias in the Birmingham District. *Quart. J. Geol. Soc.*, **89**, 53–6.

——— 1951. Permian Rocks of the Midlands. *Geol. Mag.*, **138**, 36–40.

KING, W. W. 1899. The Permian Conglomerates of the Lower Severn Basin. *Quart. J. Geol. Soc.*, **55**, 97–128.

SHOTTON, F. W. 1929. The Geology of the Country around Kenilworth (Warwickshire). *Quart. J. Geol. Soc.*, **85**, 167–222.

**Permo-Triassic**

BOSWORTH, T. O. 1912. The Keuper Marls around Charnwood. Publ. by Leicester lit. phil. Soc., Leicester and *Quart. J. Geol. Soc.*, **68**, 281–94.

BOULTON, W. S. 1951. Permian Rocks of the Midlands. *Geol. Mag.*, **88**, 36–40.

CUMMINS, W. A. 1958. Some sedimentary structures from the Lower Keuper Sandstones. *Liverpool and Manchester Geol. J.*, **2** (1), 37–43.

ELLIOT, R. E. 1961. The Stratigraphy of the Keuper Series in Southern Nottinghamshire. *Proc. Yorks. Geol. Soc.*, **33**, 197–234.

GARRETT, P. A. 1960. Nomenclature of the Keuper Series. *Nature*, **187**, 4740, 868–9.

HARRISON, W. J. 1876. On the occurrence of Rhaetic Beds in Leicester. *Quart. J. Geol. Soc.*, **32**, 212–28.

KENT, P. E. 1949. A structure Contour Map of the Surface of the Buried Pre-Permian Rocks of England and Wales. *Proc. Geol. Assoc.*, **60**, 87–104.

——— 1953. The Rhaetic Beds of the North East Midlands. *Proc. Yorks. Geol. Soc.*, **29**, 117–39.

KLEIN, G. de V. 1962. Sedimentary Structures in the Keuper Marl (Upper Triassic). *Geol. Mag.*, **99**, 137–44.

MATLEY, C. A. 1912. The Upper Keuper (or Arden) Sandstone Group and Associated Rocks of Warwickshire. *Quart. J. Geol. Soc.*, **68**, 252–80.

POOLE, E. G. and WHITEMAN, A. J. 1955. Variations in thickness of the Collyhurst Sandstone in the Manchester area. *Bull. Geol. Surv. Gt. Brit.*, No. 9, 33–41.

—— —— 1966. Geology of the Country around Nantwich and Whitchurch. *Mem. Geol. Surv.*

PUGH, W. J. 1960. Triassic Salt discoveries in the Cheshire–Shropshire Basin. *Nature*, **187**, 4734, 278–9.

RAW, F. 1934. On the Triassic and Pleistocene Surfaces developed on some Leicestershire Igneous Rocks. *Geol. Mag.*, **71**, 23–33.

RICHARDSON, L. 1912. On the Rhaetic of Warwickshire. *Geol. Mag.*, (5) **9**, 24–33.

SHERLOCK, R. L. 1926–28. A Correlation of the British Permo–Triassic Rocks. *Proc. Geol. Assoc.*, **37**, 1–72 and **39**, 49–95.

—— 1948. *The Permo-Triassic Formations*. London.

SHOTTON, F. W. 1937. The Lower Bunter Sandstones of North Worcestershire and East Shropshire. *Geol. Mag.*, **74**, 534–53.

—— 1956. Some Aspects of the New Red Desert in Britain. *Liverpool and Manchester Geol. J.*, **1**, (5), 450–65.

SHRUBSOLE, O. A. 1903. On the probable source of some of the pebbles of the Triassic Pebble Beds of South Devon and of the Midland Counties. *Quart. J. Geol. Soc.*, **59**, 311–33.

SMITH, B. 1910. The Upper Keuper Sandstones of East Nottinghamshire. *Geol. Mag.*, (5), **7**, 302–11.

STUBBLEFIELD, C. J. 1960. Nomenclature of the Keuper Series. *Nature*, **187**, 4740, 868–9.

TAYLOR, B. J., PRICE, R. H. and TROTTER, F. M. 1963. Geology of the Country around Stockport and Knutsford. *Mem. Geol. Surv.*

TONKS, L. H., JONES, R. C. B., LLOYD, W. and SHERLOCK, R. L. 1931. The Geology of Manchester. *Mem. Geol. Surv.*

WATTS, W. W. 1903. Charnwood Forest. A Buried Triassic Landscape. *Geog. J.*, **21**, 623–36.

—— 1945. Leicestershire Climate in Triassic Times. *Geol. Mag.*, **82**, 34–6.

WILLS, L. J. 1910. The Fossiliferous Lower Keuper Rocks of Worcestershire. *Proc. Geol. Assoc.*, **21**, 249–332.

—— 1947. Triassic Scorpions. Pt. 1, *Palaeont. Soc.*

**Jurassic**

ARKELL, W. J. 1933. *The Jurassic System in Great Britain*. The Clarendon Press, Oxford.

BRINKMANN, R. 1929. Statistisch-biostratigraphische Untersuchungen an Mitteljurassischen Ammoniten über Artbegriff and Stammesentwicklung. *Abh. Gesell. Wiss. Göttingen*, **13**, 1–250.

COX, A. H. and TRUEMAN, A. E. 1920. Intra-Jurassic Movements and Underground Structure of the Southern Midlands. *Geol. Mag.*, **57**, 198–208.

DEAN, W. T., DONOVAN, D. T. and HOWARTH, M. K. 1961. The Liassic Ammonite Zones and Subzones of the North-west European Province. *Bull. Brit. Mus. (Nat. Hist.) Geol.*, **4**, 435–505.

DOUGLAS, J. A. and ARKELL, W. J. 1932. The Stratigraphical Distribution of the Cornbrash, II. The North-Eastern Area. *Quart. J. Geol. Soc.*, **88**, 112–70.

FREEMAN, I. L. 1956. Variations in the Lower Zones of the Oxford Clay. *Clay Min. Bull.*, **3**, 50–61.

HALLAM, A. 1955. The Palaeontology and Stratigraphy of the Marlstone Rock-bed in Leicestershire. *Trans. Leicester lit. phil. Soc.*, **49**, 17–35.

HOLLINGWORTH, S. E. and TAYLOR, J. H. 1946. An Outline of the Geology of the Kettering District. *Proc. Geol. Assoc.*, **57**, 204–33.

KENT, P. E. 1937. The Lower Lias of South Nottinghamshire. *Proc. Geol. Assoc.*, **48**, 163–74.

——— 1940. A Short Outline of the Stratigraphy of the Lincolnshire Limestone. *Trans. Lincs. Nat. Un.*, **10**, 48–58.

KLEIN, G. de V. 1963. Intertidal zone channel deposits in Middle Jurassic Great Oolite Series, Southern England. *Nature*, **197**, 4872, 1060–2.

NEAVERSON, E. 1925. Zones of the Oxford Clay near Peterborough. *Proc. Geol. Assoc.*, **36**, 27–37.

PHILLIPS, W. 1818. *Selection of facts . . . to form an Outline of the Geology of England and Wales*. London.

RICHARDSON, L. 1923. Certain Jurassic (Aalenian–Vesulian) Strata of Southern Northamptonshire. *Proc. Geol. Assoc.*, **34**, 97–113.

——— and KENT, P. E. 1938. Weekend Field Meeting in the Kettering District. *Proc. Geol. Assoc.*, **69**, 59–76.

RUTTEN, M. G. 1956. Depositional Environment of the Oxford Clay at Woodham Clay Pit. *Geol. en Mijnb.*, **18**, 344–6.

SHARP, S. 1870–73. The Oolites of Northamptonshire. *Quart. J. Geol. Soc.*, **26**, 354–91 and **29**, 225–300.

SPATH, L. F. 1942. The Ammonite Zones of the Lias. *Geol. Mag.*, **79**, 264–8.

STRACHEY, J. 1719. A curious description of the strata observed in the coal mines of Mendip in Somersetshire . . . *Phil. Trans. R. Soc.*, **30**, 968–73.

TAYLOR, J. H. 1946. Evidence for submarine erosion in the Lincolnshire Limestone of Northamptonshire. *Proc. Geol. Assoc.*, **57**, 246–62.

——— 1951. Sedimentation problems of the Northampton Sand Ironstone. *Proc. Yorks. Geol. Soc.*, **28**, 74–85.

——— 1963. Geology of the Country around Kettering, Corby and Oundle. *Mem. Geol. Surv.*

THOMPSON, BEEBY. 1896–1905. The Junction Beds of the Upper Lias and Inferior Oolite in Northamptonshire, several papers in *J. Northants. nat. Hist. Soc.*, **9–13**.

——— 1921–8. The Northampton Sand of Northamptonshire. *J. Northants. nat. Hist. Soc.*

——— 1930. The Upper Estuarine Series of Northamptonshire and Northern Oxfordshire. *Quart. J. Geol. Soc.*, **86**, 430–62.

**Pleistocene and Recent Deposits**

BISHOP, W. W. 1958. The Pleistocene Geology and Geomorphology of Three Gaps in the Midland Jurassic Escarpment. *Phil. Trans. R. Soc.*, B, **241**, 225–305.

BOULTON, G. S. and WORSLEY, P. 1965. Late Weichselian Glaciation in the Cheshire–Shropshire Basin. *Nature*, **207**, 4998, 704–706.

BUCKMAN, S. S. 1899. The Development of Rivers; and particularly the Genesis of the Severn. *Nat. Sci.*, **14**, 273–89.

CLAYTON, K. M. 1953. Glacial Chronology of the Middle Trent Basin. *Proc. Geol. Assoc.*, **64**, 198–207.

DAVIS, W. M. 1895. On the Development of Certain English Rivers. *Geog. J.*, **5**, 127–45.

DUIGAN, S. L. 1956. Pollen-Analysis of the Nechells Interglacial Deposits, Birmingham. *Quart. J. Geol. Soc.*, **112**, 373–91.

DURY, G. H. 1951. A 400 ft Bench in South-Eastern Warwickshire. *Proc. Geol. Assoc.*, **62**, 167–73.

EASTWOOD, T., WHITEHEAD, T. H. and ROBERTSON, T. 1925. Geology of the Country around Birmingham. *Mem. Geol. Surv.*

EVANS, W. B., WILSON, A. W., TAYLOR, B. J. and PRICE, D. 1968. Geology of the Country around Macclesfield, Congleton, Crewe and Middlewich. *Mem. Geol. Surv.*

HARRISON, W. J. 1898. The Ancient Glaciers of the Midland Counties of England. *Proc. Geol. Assoc.*, **15**, 400–8.

HOLLINGWORTH, S. E. and TAYLOR, J. H. 1946. An Outline of the Geology of the Kettering District. *Proc. Geol. Assoc.*, **57**, 204–43.

KELLAWAY, G. A. and TAYLOR, J. H. 1953. Early Stages in the Physiographic Evolution of a Portion of the East Midlands. *Quart. J. Geol. Soc.*, **108** for 1952, 343–75.

KING, W. B. R. 1955. A Review of the Pleistocene Epoch in England. *Quart. J. Geol. Soc.*, **111**, 187–208.

LINTON, D. L. 1951. Midland Drainage, Some Considerations Bearing on its Origin. *Adv. Sci. Lond.*, **7**, 449–56.

MACKINTOSH, D. 1873. Observations on the more remarkable boulders of the North-West of England and the Welsh Borders. *Quart. J. Geol. Soc.*, **29**, 351–60.

PICKERING, R. 1957. The Pleistocene Geology of the South Birmingham Area. *Quart. J. Geol. Soc.*, **113**, 409–28.

POOLE, E. G. 1966. Late Weichselian Glaciation in the Cheshire–Shropshire Basin. *Nature*, **211**, 5048, 507.

────── and WHITEMAN, A. J. 1961. The Glacial Drifts of the Southern Part of the Shropshire–Cheshire Basin. *Quart. J. Geol. Soc.*, **117**, 91–130.

POSNANSKY, M. 1960. The Pleistocene Succession in the Middle Trent Basin. *Proc. Geol. Assoc.*, **71**, 285–311.

RAW, F. 1934. On the Triassic and Pleistocene Surfaces Developed on Some Leicestershire Igneous Rocks. *Geol. Mag.*, **71**, 23–31.

SHOTTON, F. W. 1953. The Pleistocene Deposits of the Area Between Coventry, Rugby, and Leamington and Their Bearing Upon the Topographical Development of the Midlands. *Phil. Trans. R. Soc.*, B, **237**, 209–60.

────── 1960. Large-Scale Patterned Ground in the Valley of Worcestershire Avon. *Geol. Mag.*, **97**, 404–8.

────── 1967 [for 1966]. The problems and contributions of methods of absolute dating within the Pleistocene Period. *Quart. J. Geol. Soc.*, **122**, 357–383.

────── 1967. Age of the Irish Sea Glaciation of the Midlands. *Nature*, **215**, 5108, 1366.

────── and STRACHAN, I. 1959. An Investigation of a Peat Moor at Rodbaston, Penkridge, Staffordshire. *Quart. J. Geol. Soc.*, **115**, 1–16.

SIMPSON, I. M. and WEST, R. G. 1958. On the Stratigraphy and Palaeobotany of a Late Pleistocene Organic Deposit at Chelford, Cheshire. *New Phytol.*, **57**, 239–250.

STEVENSON, I. P. and MITCHELL, G. H. 1955. Geology of the Country between Burton-upon-Trent, Rugeley and Uttoxeter. *Mem. Geol. Surv.*

TOMLINSON, M. E. 1925. The River Terraces of the Lower Valley of the Warwickshire Avon. *Quart. J. Geol. Soc.*, **81**, 137–69.

────── 1935. The Superficial Deposits of the Country North of Stratford-on-Avon. *Quart. J. Geol. Soc.*, **91**, 423–62.

Tomlinson, M. E. 1963. The Pleistocene Chronology of the Midlands. *Proc. Geol. Assoc.*, **74**, 187–202.

West, R. G. 1963. Problems of the British Quaternary. *Proc. Geol. Assoc.*, **74**, 147–86.

────── and Donner, J. J. 1956. The Glaciation of East Anglia and Adjoining Areas. *Quart. J. Geol. Soc.*, **112**, 69–91.

Whitehead, T. H., Dixon, E. E. L. and others. 1928. Country between Wolverhampton and Oakengates. *Mem. Geol. Surv.*

Wills, L. J. 1912. Late-Glacial and Post-Glacial Changes in the Lower Dee Valley. *Quart. J. Geol. Soc.*, **68**, 180–98.

────── 1924. The Development of the Severn Valley in the Neighbourhood of Ironbridge and Bridgnorth. *Quart. J. Geol. Soc.*, **80**, 274–314.

────── 1937. The Pleistocene History of the West Midlands. *Rep. Brit. Assoc.*, 1937, 71–94.

────── 1938. The Pleistocene Development of the Severn from Bridgnorth to the Sea. *Quart. J. Geol. Soc.*, **94**, 161–242.

Yates, E. M. and Moseley, F. 1957. Glacial Lakes and Spillways in the Vicinity of Madeley, North Staffordshire. *Quart. J. Geol. Soc.*, **113**, 409–28.

### Structure

Carder, D. S. 1945. Seismic investigations in the Boulder Dam area, 1940–1941 and the influence of reservoir loading on earthquake activity. *Seismological Soc. America Bull.*, **35**, 175–92.

Davison, C. 1924. *A history of British Earthquakes*. Cambridge.

Dollar, A. T. J. 1957. The Midlands Earthquake of February 11th, 1957. *Nature*, **179**, 507.

Fuller, N. L. 1912. The New Madrid Earthquake. *U.S. Geol. Surv. Bull.*, **494**.

Hollingworth, S. E., Taylor, J. H. and Kellaway, G. A. 1944. Large-scale superficial structures in the Northampton Ironstone Field. *Quart. J. Geol. Soc.*, **100**, 1–44.

Kellaway, G. A. and Taylor, J. H. 1953. Early stages in the Physiographic Evolution of a portion of the East Midlands. *Quart. J. Geol. Soc.*, **108**, 343–76.

Lawson, A. C. (Editor). 1908. The Californian Earthquake of April 18, 1906. *Report of the State Earthquake Investigation Committee*, **1**.

McGinnis, Lyle D. 1963. Earthquakes and Crustal Movement as Related to Water Load in the Mississippi Valley Region. *Illinois State Geological Survey Circular* 344, 1–20.

Oldham, R. D. 1899. Report on the great earthquake of 12th June, 1897. *Mem. Geol. Surv. India*, **29**, 1 379.

Shotton, F. W. 1965. Normal faulting in British Pleistocene deposits. *Quart. J. Geol. Soc.*, **121**, 419–34.

Taylor, J. H. 1947. Evidence of submarine erosion in the Lincolnshire Limestone of Northamptonshire. *Proc. Geol. Assoc.*, **57** [for 1946], 246–62.

────── 1963. The Geology of the Country around Kettering, Corby and Oundle. *Mem. Geol. Surv.*

# Index

Abberley, 58
──── Breccia, 57
──── Hills, 26, 48, 59
Abbey Shales, 12
Abdon, 25
──── Limestones, 22, 23, **25**
Acton Round, 24
Adderley, 98
Age determinations, of Charnian, 9; of Mountsorrel Granodiorite, 27; of Whitwick Dolerite, 56
Aggborough, near Kidderminster, 66
Alcester, 70
Alderley Edge, 73
ALLEN, J. R. L., 6, 23
Allesley Conglomerate, 58
Alluvium, 101
Amington Hall, borehole at, 32
Anglo-Welsh area, 23, 24
Apedale Fault, 47
Arden Sandstone, 70
Arid conditions, 57, 61, 71
Arley Anticline, 47
──── Conglomerate, 58
──── Fault, 47, 111
Armorican Front, 111
Ashby Anticline, 47
Ashby-de-la-Zouch, 32
Ashow Breccia, 57
──── Group, 57, 58, 59
Astbury, 28, 31
Atherstone, 22, 47, 110
AVELINE, W. T., 4
Avon (Warwickshire), River, 1; Second Terrace of, 94, 99; Third Terrace of, 95; Fourth Terrace of, 95; terraces (general), 101
Aymestry (Sedgley) Limestone, 15, 20

Baginton sand, 95
──── -Lillington gravels, 95
Bagot's Park, borehole at, 71
BALL, H. W., 20, 22
Banbury, 72, 77, 78, 105
Bardon, 108
──── Hill, 7

Barnack, 84, 107, 119
──── Stone, 84, 107
Barnstone, near Nottingham, 104
Barnt Green, 6, 7, 13, 18
Barr, 48, 104
──── Beacon Beds, 64
Barrow, 17, 20
──── Hill (near Ashby-de-la-Zouch), 32
──── ──── (near Burton-upon-Trent), 72
──── ──── (near Pensnett), basalt, 55
Basement Beds, Wenlock Limestone, 19
Bawdon Castle, 8
Beacon Hill, Charnwood Forest, viii
──── ──── Beds, viii, 7
Bedworth, 10, 47
Beechwood Conglomerate, 58
Bewdley, 48
Bickerton–Peckforton Fault, 73
Biddulph, 37
Binley, 47
Binton, 77
Birmingham, 4, 61, 94
──── University, Geological Department of, 5, 20
BISHOP, W. W., 96, 101
Blackband Group, **53**
──── ironstones, 42, 53, 105
Blackbrook Beds, 7
──── Series, 7
Blue Hole Intrusive Series, 6
Blue Lias, 76–7, 104
BONNEY, T. G., 5
Boothorpe Fault, 47
Boreholes at, Amington Hall, 32; Bagot's Park, 71; Cambridge, 31; Chartley, 32, 71; Claverley, 55; Ellistown, 32; Gayton, 22, 27, 31, 62; Glinton, 6, 7; Great Barr, 17; Great Oxendon, 6, 7; Great Paxton, 110; Hathern, 34; Kettering, 28; Leicester, 14; Long Clawson, 34, 39; Market Bosworth, 14; Market Drayton, 32; Merevale (Atherstone), 12, 38; Mickleton,

133

77; Nechells, 12; Northampton, 28, 30; North Creake, 6; Nuneaton, 14; Orton, 6, 7, 61; Rugeley, 38; Sapcote, 14; Shuttington Fields, 14; Sproxton, 6, 7, 39; Statfold, 32; Stockshouse Farm, 30, 32, 38; Walsall, 13, 18; Whittington Heath, 22, 27, 30, 32, 38; Widmerpool, 34, 39; Wilkesley, 73, 77; Wyboston, 111
BOSWORTH, T. O., 5
Boulder clay, 89, 99
BOULTON, G. S., 99
BOULTON, W. S., 5
Bowhills, 59
——— Group, 57, **58**
Brackley, 83, 85
Bradgate, 8
Brand Series, 7
Brazil Wood, 9, 27
Breccia Group, 57, **58–9**
Bredon Hill, 94, 115
Breedon, 108
——— Cloud, 28, 32
——— on the Hill, 32
Brewin's Bridge, near Netherton, 26
Brick Clays, 105
Bridgnorth, 20, 48, 61, 63, 66, 69, 112
Brine, 71, **106–7**
BRINKMANN, R., 88
Brixworth, 79
BRODIE, P. B., 4
Bromsgrove, 69, 108
Brown Clee Hill, 24, 25, 50; basalt at, 55
——— ——— Syncline, 111
Brownstones, 25
Bubbenhall clay, 94
Budleighense River, 66
Building Stone, 107
Buildwas, 98
Bulkeley, viii
Bunter, **63–6**
——— Pebble Beds, 32, 47, **64–6**, 106
Burton-upon-Trent, 64, 109
BUTLER, A. J., 18


Calcareous Conglomerate Group, **57, 58**
Caldecote, near Nuneaton, 6, 7
——— Volcanic Series, **6, 9**

Calke, 32
Cambers, 114
Cambrian, 6, **10–4**, 31, 47, 51
——— quartzite, 6, 7, 9, 17, 108; pebbles of, 58
Cambridge, 30; borehole at, 31
Cannock Chase, 41, 51, 64
Carboniferous, **28–56**
——— Limestone (Series), 27, **28–34**, 39; 'basin' facies of, 28, 32; basalt in, 31–2, 108; evaporites in, 34; 'massif' facies of, 28, 34; pebbles of, 58, 66; shorelines of, 30; Zones and Stages of, 30
CARDER, D. S., 117
Castle Hill, 59
Cement, 104
Chalky Boulder Clay, ix, 94, 95, 96
*Chara* marl, 101
Charnian, 6, **7–9**, 27; clastic rocks, 7; fossils in, 8; igneous rocks, 8, 108; 'markfieldite' ('syenite'), 6, 8, 9, 27, 108; 'porphyroids', 7, 8, 108
Charnwood Forest, 1, 5, 6, 7, 9, 28, 68, 71, 108, 109, 111, 117
——— ———, structure of, 8
Chartley, borehole at, 32, 71
Chatsworth Grit, 37
Cheadle Coalfield, 36, **45**, 47, 53
Chelford, 94, 108
——— Sand, 98–9
Chellaston, near Derby, 107
Cherwell, 96
Cheshire Basin, 113
——— Plain, 1
Chester, 68
Church Stretton Fault Complex, 111
Claverley, 59; borehole at, 55
Clee Group, 22, 23, 24, **25**
——— Hills, 1, 22. 23, 24, 25, 27 45, 55
——— ——— Coalfield, **48**
Clent Breccias, 57, 59
——— Hills, 1, 59, 96, 100
Clipsham Stone, 107
Clive, 73
Coal, 4, 41; production of, 104
——— Measures, 12, 13, 14, 18, 20, 24, 26, 28, 31, 38, **39–56**; classification of, 43–4; fauna and flora of, 42–6; igneous rocks in, 55–6, 108; *Lingula* bands in, 42; Marine Bands in, 42, 43, 44, 45
Coalbournbrook, 27

## Index

Coalbrookdale, 41, 48, 50, 54, 106
——— Coalfield, 6, 15, 20, 21, 26, 28, 31, 45, **48**, 52, 54, 57
Coalfields of Central England, **45-50**
Coalport Beds, 54
Coaly Bed, Lower Estuarine Series, 83
COBBOLD, E. S., 5
Colesheath, 108
Collyhurst Sandstone, 64
Collyweston Slate, 84, 107
Colsterworth, 83
Comley Sandstone, 14
Congleton, 108
——— Edge, 37
——— Sand, 98, 108
Conglomerates, 58
Continental beds, 23, 61
Corby, 83, 105
Corley Beds, 59
——— Conglomerate, **58**
Cornbrash, **87**
Cornbrook, 55
——— Sandstone, 31, **37-8**, 50
Cornstones, 23, 24, 25, 42
Corve Dale, 24, 25
Cosgrove, viii
Costock, 76
Cotham Beds, **73**
Coventry, 4, 58, 95
Croft, 9, 108
'Crog-balls', Wenlock Limestone, 19
CROOKALL, R., 43, 44
*crossi* Bed, 84
Cryoturbation structures, 116
CUMMINS, W. A., 69


Daventry, 77, 78
DAVISON, C., 118, 119
Dee, River, 1
Derby, 32, 119
Dimminsdale, 32
DINELEY, D. L., 22, 23
Dip-and-fault structure, 114-15
Ditton Series, viii, 22, 23, 24, **25**, 26
DIXON, E. E. L., 97
DOLLAR, A. T. J., 119
DONNER, J. J., 94
Doseley, 32, 108
Dosthill, 10, 12, 14, 38, 47
Downton Castle Sandstone, viii, **22**, 23, **24**, 26

Downton Series, 17, 22, 23, **24**, 25, 26
Droitwich, 76, 100, 109
Duddington Fault, 118-9
Dudley, viii, 17, 18, 19, 20, 41, 48, 104; Thick Coal at, 39, 41, 52
——— Castle Hill, 17, 18, 20
——— Museum, 20
DUIGAN, S. L., 94
Dune Sandstone Group, 63
Dunsmore Gravel, 95-6
DURY, G. M., 96


Eakring, 39
Earthquakes, 118-9
Eastern Boundary Fault, 59
——— Drift, 90, 94
EASTWOOD, T., 56, 64, 97
Eccleshall, 98
Economic Geology, **104-9**
Edge Hill, 77, 78, 107
EDMONDS, E. A., 101
EDMUNDS, F. H., iii
EDWARDS, W., 44
Ellesmere Moraine belt, 96
Ellistown, dolerite at, 56; borehole **at**, 32
Enderby, 9, 108
Enville, 58
——— Beds, 47, **56-9**; Uriconian pebbles in, 7
——— Breccias, 57, 59
——— Fault, 48
'Espleys', 42, 53, 54
Etruria Marl (Group), 41, 42, 51, 52, **53-4**, 105
Eurypterid Sandstones, 26
EVANS, A. M., 8
EVANS, W. B., iii, 92, 98, 99
Evaporites, 34, 61, 71; anhydrite, 34; gypsum, 71; salt, 71
EVERITT, C. W. F., 55
Evesham, 73, 76, 100
Exhall, 47
Eye Brook, 115


Fair Oak Colliery, 32
FALCON, N. L., 28, 32, 56
FAREY, J., 4
Farlow Sandstone Series, 22, 23, **27**, 111

Fauld, near Tutbury, 107
Felsitic Agglomerate, 7, 8
Fen deposits, 101
Fenny Compton spillway, 96
Fireclay, 4, 105
First Welsh Glaciation, 92, 95
Fluvioglacial sand and gravel, 99
FORD, T. D., 8
Fotheringhay, 85
Four Ashes, near Wolverhampton, 99
FOX-STRANGWAYS, C., 5
FREEMAN, I. L., 88
Frodsham Beds, 68
FULLER, N. L., 117


Ganisters, 108
Gawsworth Sand, 98–9
Gayton, 30; borehole at, 22, 27, 31, 62
GEORGE, T. N., 38
Gibbet Hill Conglomerate, 58
————— Group, 57, 58
Glacial erratics, 89, 90
Glacial lakes: Buildwas, 97, 98; Cannock Chase, 96; Coalbrookdale, 96, 97, 98; Harrison, 95–6; Lapworth, 96, 98, 99; Madeley, 98; Newport, 96, 97, 98; Oakengates, 96; Shropshire–Cheshire Basin, 98
Glen Parva Brickworks, 72
—————, River, 1
Glinton, borehole at, 6, 7
Gnossall, 98
Gornal, 26
————— Sandstone, 26
Grace Dieu, 32
————— Manor, 7
Grantham, 28, 77, 80, 83, 84
Great Barr, 17, 18; borehole at, 17
Great Oolite Clay (Blisworth Clay), **86–7**
————— Limestone (Blisworth Limestone), viii, **86**, 104
————— Series, **85–7**
————— Oxendon, borehole at, 6, 7, 62
————— Paxton, borehole at, 110
————— Ponton, 104
————— Weldon, ix
Gretton, near Uppingham, 109
Grey Downton (Temeside) Group, 22

Grimley, near Thringstone, 8
Grinshill, 73, 108
Groby, 8, 108
Gulls, 114, 117
Gypsum, 4, 71, 107


Haffield Breccia, 57
Halesowen, 18, 52
————— Beds, 41, 53, 54
Hammercliffe, near Copt Oak, 8
Hamstead, 59
Hanging Rocks Conglomerate, 7
Hanley, 53
Hartshill, 6, 108
————— Quartzite, 12, 13, 110
Hathern, borehole at, 34
Head, 100
Heightington, 22, 25, 26, 48
Henley-in-Arden, 76
Hercynian earth-movements, 45, 60
Highley Beds, 48, 54
High Sharpley, near Whitwick, 8
Hilton Main Colliery, 32
HIND, W., 43
Hollington Quarries, near Uttoxeter, 108
HOLLINGWORTH, S. E., 80, 94, 115, 117
Holowell, 115
Hopwas Breccia, 64
HORTON, A., 52
HOWELL, H. H., 4
HULL, E., 4
Hurst Hill, 17, 18
*Hyolithes* Limestone, Hartshill Quartzite, 12


Ignimbrite, 6
ILLING, V. C., 5, 12
Inferior Oolite Series, **79–84**
Intermontane basins, 60
Intra-Carboniferous Movements, 111–2
Irish Sea Drift, 90
Ironbridge, 31, 48, 98
————— col, 97
————— Gorge, 97, 98, 100
Iron ore, 4, 78–9, 80, 105
IVIMEY-COOK, H. C., iii

## Index

JONES, D. G., 38, 50
JONGMANS, W. J., 43
JUKES, B., 4
Jurassic, **74–88**

Keele Group, 42, **54–5**
KELLAWAY, G. A., iii, 102, 115, 116
Kellaways Beds, **87–8**
Kenilworth, 58
——— Breccias, 57
——— Sandstones, 57, 58
KENT, P. E., 28, 32, 56
Kettering, 28, 83, 85, 86, 105, 108; borehole at, 28
Ketton, 104
——— Stone, 107
Keuper, **66–73**
——— Basement Beds, 68
——— Building Stone Group, 68, 108
——— Conglomerate, viii
——— Marl, **69–72**, 105, 109
——— Passage Beds, viii
——— Saliferous Beds, 71
——— Sandstone, **68–9**, 109
——— ——— Conglomerate 68
Kidderminster, 64, 108
KIDSTON, R., 38, 43, 44
Kilsby, 117
Kinchley, 9, 27
Kinderscout Grit, 37
——— ——— Group, 37
KING, W. W., 5, 24, 25, 26
Kingscliffe, 119
Kinlet, basalt at, 55
——— Beds, 48
KLEIN, G. de V., 70, 86
Knowle, 72

Landslips, 115–6
Landywood, basalt at, 55–6
LAPWORTH, C., 5, 12
Lateritization, 61
LAWSON, A. C., 117
Lea Hall Colliery, 48
Leamington, 73, 95, 109
Leicester, 72, 73, 95, 118; borehole at, 14
Leicestershire Coalfield, 14, 28, 32, 38, 45, **47**, 51, 52, 53, 55

Lias, **75–9**; ammonite zones of, 75
Lichfield, 32, 57
Lickey Hills, 7, 10, 13, 15, 17, 48
——— Quartzite, 13, 108; pebbles of, 66
Lightmoor, 97
Lilleshall, 6, 7, 10, 14, 28, 31, 48, 100, 105
——— Fault, 48
Limestone, 104
Lincoln, 105
Lincolnshire Limestone, ix, **83–4**, 104, 107, 109, 115, 119
Linley, 17, 20, 48
LINTON, D. L., 102
Little Wenlock, 28
Llandovery Sandstone, pebbles of, 66
Llanvirn, 110
London Platform, 61, 68, 74, 113
Long Clawson, borehole at, 34, 39
Longmyndian, 6
Loughborough, 119
Lower Boulder Clay, 92, 96
——— Carboniferous, 23, 27, **28–34**
——— Coal Measures, 39, 41, 42, 45, 47, 48, **50–1**, 56; rhythmic sedimentation in, 41
——— Estuarine Series, ix, 82, **83**, 84, 108
——— Lias, **76–7**
——— Ludlow Shales, 15, **20**
——— Mottled Sandstone, **63–4**
——— Old Red Sandstone, 22, **24–7**
——— Quarried Limestone, Wenlock Limestone, 19
——— Rhaetic (Westbury Beds), **72–3**
——— Wenlock Limestone, 19
Ludlow Bone Bed, 22, 23, **24**, 25, 26
Ludlow–Ledwyche Anticline, 111
Ludlow Series, 15, 17, **20–1**
Lydebrook Sandstone, 31
Lye, 17, 20, 21, 26, 32,


Macclesfield, 98–9
Main Irish Sea Glaciation, 92, 96
Malpas Sandstone, 69
Malvern–Abberley Axis, 25, 48
Mamble Coalfield, 25, 26, 48
Manchester, 98

Manchester Marl, 64
Maplewell Series, 7, 8
Marine Bands: Alton, 51; *Anthracoceras cambriense* 45; *A. vanderbeckei*, 45, 51, 52; Bagworth, 52; Bay, 51, 53; Blackstone, 52; Chance Pennystone, 52; Charles, 52, 54; Eymore Farm, 52; *Gastrioceras listeri*, 50; *G. subcrenatum*, 45, 50, 51; Gin Mine, 51; Longton Hall, 52; Molyneux, 52; Nuneaton, 52, 54; Pennystone, 52; Priorsfield, 52; Seven-Feet Banbury, 50, 51; Seven-Feet, 52; Sub-Brooch, 52
Market Bosworth, 95; borehole at, 14
——— Drayton, borehole at, 32
——— Weighton Axis, 74
Markfield, 8, 108
Marlstone Rock Bed, **78–9**, 105, 107, 109
MARSHALL, C. E., 55
MCGINNIS, L. D., 117
Melbourne, 38, 109, 115
Melton Mowbray, 32, 56, 79, 105
——— ——— Anticline, 74
MENEISY, M. Y., 9, 27, 56
Mercian Highlands, 30, 34, 39, 57, 58, 60, 66
Merevale, 10; borehole at, 12, 38
——— Shales, 12
Merry Lees Colliery, 14
Mickleton, borehole at, 77
Middle Coal Measures, 39, 41, 42, 45, 47, 48, **51–3**, 56, 105, 106; rhythmic sedimentation in, 41
——— Grit Group, 37
——— Lias, **77–9**
——— ——— Clays and Silts, 78
——— Sands, 96, 97
Midland Barrier, 39, 41
——— Block, 15, 22, 23, 111, 113
MILLER, J. A., 9, 27, 56
Millstone Grit (Series), 28, 31, 32, **34–9**, 41, 45, 108; breccia bed at base of, 38; Marine Bands in, 34; rhythmic sedimentation in, 34, 41; Stages, Groups and Zones of, 36
Mineralization, 73
MITCHELL, G. H., 100
Moira Breccia, 63, 64
Monks Park Shales, 12
Moor Wood Flags and Shales, 12
Moreton-in-Marsh, 95, 96

MOSELEY, F., 98
Moulding Sand, 108
Mountsorrel, 71, 100
——— Granodiorite, 9, **27**, 108, 111
Much Wenlock, 105
MURCHISON, R. I., 4
Muschelkalk, 69

Nantwich, 96, 98
Narborough, 9, 108, 117
Nechells, 91; borehole at, 12
——— Breccia, 12, 57
Needwood Basin, 113
Neen Sollars, 17, 20, 21
Nene, River, 1
Netherton, 17, 26
——— Anticline, 20, 26
Newark-on-Trent, 28, 56
Newcastle-under-Lyme Group, 41, **54**
Newer Drift, 95, 96–7
Newhurst, 108
New Red Sandstone, 45, **60–72**
Nodular Beds, Wenlock Limestone, 18, 19
North Creake, borehole at, 6
——— Staffordshire Coalfields, **45**, 50, 52, 53
Northampton, 28, 80, 83, 85, 119; borehole at, 28, 30
——— Sand, **80–3**, 107, 109
——— Ironstone, ix, 105, 115
Northern Drift, 90, 95
Nottingham, 28, 32
Nuneaton, 1, 6, 9, 10, 12, 13, 14, 96, 108, 110, 111; borehole at, 14

Oakham, 77, 79, 83, 118
OAKLEY, K. P., iii
Oil, 106
Oldbury, 20
——— Shales, 12
Older Drift, 92, 95, 96–7
OLDHAM, R. D., 117
Old Red Sandstone, 17, **22–7**, 42, 48, 51, 55
Onibury, viii
Ordovician, 14, 15; microdiorites, 6
Orton, 86; borehole at, 6, 7, 61

Osgathorpe, 32
Oundle, 83, 85, 109
Outwoods Shales, 12
OWEN, T. R., 38, 50
Oxford Clay, **88**, 106

Park Hill, 20
Passage Beds, Wenlock Limestone, 19
Patterned ground, 100
Peckforton, 66, 68
Peldar Tor, 8
Penkridge, 108
Periglacial Deposits, **99–100**
Permian, **63**, 64
Permo-Carboniferous, **56–9**
———— Movements, **111–3**
Permo-Triassic, 41, 56, **60**–73
———— Sandstone, 64
Peterborough, 106
PHILLIPS, W., 75
PICKERING, R., 96
Pleistocene, **89–100**; nomenclature and classification of, 90–2; sequence in, 92–9
———— Structures, 5, **114–6**
PLOT, R., 4
Plungar, 39, 56, 106
POCOCK, R. W., 31, 55
PODMORE, J. S., 27
Polesworth River, 64, 65
POOLE, E. G., 63, 69, 73, 92, 93, 96, 97, 98, 99
POSNANSKY, M., 100, 102
Post-Glacial and Recent Deposits, **101**
Post-Triassic Movements, 113
Potteries Coalfield, 36, 37, **45**, 53
———— Syncline, 36, 112
Pottery Clays, 105
Pouk Hill, basalt at, 55, 56
Pre-Cambrian, **6–9**, 39; pebbles of, 59
Pre-Old Red Sandstone Movements, 111
Prees Syncline, 72, 74, 113
Proto-Avon valley, 95, 96
'*Psammosteus*' Limestones Group, 22, 23, 24, **25**, 26
Purley Shales, 12

Quaternary Earth Movements, **117–9**

Radiocarbon dating, 94, 99
RAW, F., 5, 71, 100
Red Downton (Ledbury) Group, 22, 23, **24**, 26, 27
———— Rock Fault, 31, 45
Refractory Clays, 105
———— Sands, 108
Rhaetic, 72–3
Rhythmic deposition, 23, 25, 34, 41, 53, 54, 61, 64, 71
River Drainage, changes in, **102–3**
———— Terraces, 101
Road-stones, 108
Rock, 26
———— End, 37
———— Salt, 4, 71, **106–7**
Rodbaston, near Penkridge, 101
Rough Rock, 37
———— ———— Group, 37
Rowley Hills, basalt at ('Rowley Rag'), 55, 56
———— Regis, 1, 17, 18, 108
Rubery, 17
———— Sandstone, 15, 17
———— Shale, 15, 17
Rugby, 76, 77, 95, 96, 104
Rugeley, 48; borehole at, 38
RUSHTON, A. W. A., iii, 12

Sags, 115
St. George's Land, 30, 34, 39
Saltwells, 20, 21
Sand and Gravel, ix, 106
Sandiway, 99
Sandonbank, 48
Sapcote, 9; borehole at, 14
Screveton, 56
Second Welsh Glaciation, 92, 94
Sedgley, 20, 21, 26, 48, 104
———— and Dudley Anticline, 17
———— Beacon, 20
Sedgley–Lickey Axis, 74
Seismic activity, 117–8
Senni Beds, 25
Severn Basin, 61, 113
————, River, 1, 48, 96, 97, 98; Kidderminster Terrace of, 95, 98; Main Terrace of, 94, 95, 98, 99, 100; Worcester Terrace of, 95, 101; other terraces of, 101
Shaffalong Coalfield, 45
Shatterford, basalt at, 55

Shawell, gravel pit at, ix
Shell marl, 101
SHERLOCK, R. L., 72
SHOTTON, F. W., 5, 58, 94, 95, 99, 100, 101, 102, 117
Shrewley, 70
Shrewsbury, 98
Shropshire–Cheshire Basin, 92, 96–9; drift sequence in, 97
Shuttington Fields, borehole at, 14
Silica Sands, 108
Silurian, **15**–**21**, 23, 26, 48; 'basin' facies of, 15; pebbles of, 58; 'shelf' facies of, 15
SIMPSON, I. M., 94
Skerries, 70
Slate-agglomerate, 7, 8
SMITH, J. D. D., 12
SMITH, W., 4
Snelston Common, 32
South Derbyshire Coalfield, 45, **47**, 51, 52, 54, 55
——— Staffordshire Coalfield, 15, 18, 20, 22, 26, 27, 28, 32, 38, 41, 45, **48**, 50, 52, 53, 54, 55; pre-Carboniferous floor of, 17, 26, 48
SPINK, K., 52
Sproxton, 56; borehole at, 6, 7, 39
SQUIRRELL, H. C., 20
Stafford, 53, 105, 119
Stamford, 83, 84, 85, 105, 118, 119
Stanley Grit, 37
Statfold, borehole at, 32
STEVENSON, I. P., 100
Stockingford Shales, 12
Stockshouse Farm, borehole at, 30, 32, 38
Stoke-on-Trent, 1, 4, 37
Stoke Prior, near Droitwich, 71
Stone, 53
Stony Stratford, 83
Stourbridge, 17, 69
Stourport, 26
Stow on the Wold, 77
STRACHAN, I., 101
STRACHEY, J., 75
Stratford-upon-Avon, 57, 74, 76
Structure, **110**–**9**
STUBBLEFIELD, C. J., 39, 44, 45
SWINNERTON, H. H., 5
Swithland, 108
——— Reservoir, 27
——— Slates, 7, 27
'Symon Fault', 41, 48, 53, 54, 112

Table of Formations, 2–3
Tame Valley, 94, 96–9
Tamworth, 47
TARLO, L. B. H., 23
TAYLOR, B. J., iii, 70
TAYLOR, J. H., 80, 83, 84, 86, 94, 102, 115, 116, 119
TAYLOR, K., 27
Tea Green Marl, 70, 71
Temeside Shales, 22, 23, **24**, 26
Thames, River, Wolvercote Terrace of, 96
THOMPSON, B., 5
Thringstone Fault, 14, 47
Ticknall, 32
Tile Hill Group, 57, **58**
Till, 89
Tinwell Fault, 118–9
Tipton, 20
Titterstone Clee Hill, viii, 24, 27, 28, 30, 37, 50, 100, 108, 111; basalt at, 55
TOMLINSON, M. E., 101
Tong, 108
TONKS, L. H., 63
Tournaisian, 28, **30**–**1**, 38
Towcester, 80, 83, 85, 86, 87, 105
Trachose Grit and Conglomerate, 7
Transition Bed, 79
Tremadoc Series, 12, 14
Trent, River, 1; Beeston Terrace of, 95; Hilton Terrace of, 95
Triassic, viii, 14, 17, 27, 45, 47, **63**–**73**; Llandovery pebbles in, 18
——— Basins, 61, 69
Trimpley, 22, 25, 48
——— Fish Zone, 25
TROTTER, F. M., 39, 44, 45
Troughstone Hill, 37
TRUEMAN, A. E., 43
Tufa, 101
Turner's Hill, 20, 21, 26
——— ——— Beds, 26

———

Upper Boulder Clay Glaciation, 92, 96–8
——— Carboniferous, 28
——— Coal Measures, 39, 41, 42, 43, 45, 47, 48, **53**–**5**, 105, 106; Llandovery Sandstone, pebbles of, 18; rhythmical sedimentation in, 42, 53

## Index

Upper Estuarine Series, ix, **85–6**, 104
———— Lias, **79**
———— Llandovery Series, 13, 15, **17–8**
———— Ludlow Shales, 15, **20–1**, 22, 24, 26
———— Mottled Sandstone, 66
———— Old Red Sandstone, 22, 23, **27**, 32, 47
———— Quarried Limestone, Wenlock Limestone, 19
———— Rhaetic (Cotham Beds), **73**
———— Wenlock Limestone, 19
Uppingham, 79
Uriconian, **6**, 7, 13, 14; vulcanicity of, 6

Vale of Belvoir, 63
Valley bulges, 115
Variable Beds, 80
VAUGHAN, A., 30
Viséan, 28, 30, **31–4**

Walsall, 17, 18, 19, 20, 104; borehole at, 13, 18
Waltham on the Wolds, 104
Warshill, 59
Warwick, 47
Warwickshire Coalfield, 1, 10, 14, 22, 26, 27, 28, 32, 38, 45, **47**, 51, 52, 54, 55, 57, 58, 111
'Wash-outs', 42
Waterstones, 68
Water supply, 109
WATTS, W. W., 5, 7, 27
Weaver, River, 1
Wednesfield, basalt at, 55, 56
WEIR, J., 43
Weldon, 84
———— Stone, ix, 107
Welland, River, 1
Wellingborough, 83, 105
Welsh Geosyncline, 14, 15, 22
———— Uplands, 57
Wem Fault, 113
Wem–Audlem area, 72, 74, 77
Wenlock (Dudley) Limestone, viii, 15, **18–9**, 105; pebbles of, 58; reef limestones in, 15, 19
———— Series, 15, 17, **18–20**

Wenlock Shales, 15, **18**, 20
WEST, R. G., 92, 94
West Bromwich, 13; shaft at, 13, 18
Westbury Beds, **72**
Western Anticline, 112
———— Drift, 90, 94
WHITE, D. E., 12
WHITE, E. I., 22, 23, 24
WHITEHEAD, T. H., 56, 97
White Lias, **76**
WHITEMAN, A. J., 63, 69, 73, 92, 93, 96, 97, 98, 99
Whitemoor Brickworks, near Kenilworth, 59
Whittington Heath, borehole at, 22, 27, 30, 32, 38
Whitwick, dolerite at, 55, 56
Whixall Moor, near Ellesmere, 101
Widmerpool, borehole at, 34, 39
———— Gulf, 28, 32, 38, **39**
Wigston, 76
Wilkesley, borehole at, 73, 77
Willey, 17, 20
WILLS, L. J., 5, 15, 17, 57, 59, 64, 66, 69, 90, 92, 95, 96, 97, 98, 101, 102
WILSON, A. A., iii
Wind-etched stones, 100
Wingfield Flags, 51
Winsford, 71, 106
Wittering, 118
Wollescote, 20, 26, 32
Wolston Series, 95–6
Wolverhampton, 97, 99, 108
Woodhouse and Bradgate Beds, 7, 8
———— Eaves, 108
Woolhope (Barr) Limestone, 15, **18**
WORSLEY, P., 99
Wrekin, The, 6
Wren's Nest Hill, 17, 18, 20
Wrexham–Whitchurch Middle Sands morainic ridge, 98
Wyboston, borehole at, 111
Wyre Forest, 20, 41, 54, 57, 58, 59
———— ———— Coalfield, 22, 25, 26, 41, 45, **48**, 51, 52, 54, 55, 57, 59

YATES, E. M., 98
YATES, J., 4
Yorkshire–East Midlands Coalfield, 55, 56

Zechstein Sea, 60

Printed in the United Kingdom for Her Majesty's Stationery Office
Dd. 238949.   8/87.   C50.   398.   CCN 12521